Residential

Load

Calculation

Manual J

Air Conditioning Contractors of America

Create New Opportunities for Business Success Using ACCA Products and Services.

Technical Manuals

ACCA's technical manuals set the industry standard for state-of-the-art techniques, and provide the information and data that HVACR contractors need for proper system design, equipment selection and installation procedures. Now available through ACCA, the latest editions of ACCA's residential and commercial manuals include:

- **Commercial Applications, Systems and Equipment**
  Manual CS

- **Residential Duct Systems**
  Manual D

- **Heat Pump Systems: Principles and Applications**
  Manual H

- **Residential Load Calculation**
  Manual J

- **Commercial Load Calculation**
  Manual N

- **Psychrometrics: Theory and Applications**
  Manual P

- **Commercial Low Pressure, Low Velocity Duct System Design**
  Manual Q

- **Residential Equipment Selection**
  Manual S

- **Air Distribution Basics for Residential and Small Commercial Buildings**
  Manual T

- **Installation Techniques for Perimeter Heating and Cooling**
  Manual 4

- **Technical Topics: Understanding the Friction Chart**
  TT-102

Ordering ACCA products is easy!

Order by fax 24 hours a day at (301) 604-0158. Call ACCA Distribution Center today at 888-290-2220 for orders or more information about other ACCA products and services.

Air Conditioning Contractors of America
2800 Shirlington Road, Suite 300
Arlington, VA 22206
Phone: (703) 575-4477 • Fax: (703) 575-4449
Internet: www.acca.org

ACKNOWLEDGEMENTS
MANUAL J
SEVENTH EDITION

This manual was prepared by the
Air Conditioning Contractors of America

AUTHOR:

Hank Rutkowski, P.E.
ACCA Technical Director

Review and Technical Assistance:

Rod Beever
Product Manager, Borg-Warner Environmental Systems, Inc.,
(York, PA)

Don Cochell
Technical Advisor, Detroit Edison
(Detroit, MI)

R. Lee Culpepper
Program Manager, Tennessee Vallee Authority
(Chattanooga, TN)

Stephen D. Kennedy
Southern Company Services
(Atlanta, GA)

Nance Lovvorn
Technical Advisor, Alabama Power Company
(Birmingham, AL)

Thomas McGarry
Residential Marketing, Training Director
Rochester Gas & Electric (Rochester, NY)

Barney Menditch, P.E.
President of General Heating Engineering Co., Inc.
(Capitol Heights, MD)

Alfred A. Piff
Manual Committee Chairman
President of A/C Contracting Company, Inc.
(Mobile, AL)

Robert Vickery
Southern Services Co.
(Atlanta, GA)

Ron Yingling
NAHB Research Foundation
(Rockville, MD)

Reader Response

ACCA is dedicated to providing its members and users of all ACCA manuals with accurate, up-to-date and useful information. If you believe that any of the information contained in this manual is incomplete or inaccurate, if you have suggestions for improving the manual, or if you have general comments, we'd like to hear from you. Please write your comments below and return this form to:

Air Conditioning Contractors of America
2800 Shirlington Road, Suite 300
Arlington, VA 22206
Phone: (703) 575-4477 • Fax: (703) 575-4449
Internet: www.acca.org

Comments on ACCA's Manual J:

_____

_____

_____

_____

_____

_____

Name _____

Address _____

_____

Phone _____

Manual J
Seventh Edition
Residential Load Calculation

Table of Contents

COPYRIGHT 1986:
Reprinted 2019
ISBN: 1-892765-01-2

List of Figures

List of Tables and Calculation Procedures

How To Use This Manual

The first six sections of this manual outline a simple, accurate procedure which can be used to estimate the heat loss and heat gain for conventional residential structures. These sections (and Tables 1-9) form a module that is specifically intended to provide step-by-step instructions on how to perform the Manual J calculations.

Those who want to know more about the underlying assumptions and physical principles of the calculation procedures should look at Section 7. This section defines the limitations of the procedure. It will be particularly useful to those whose applications fall outside of the range of conventional construction practices. It will also be useful to those who intend to teach others on the residential heat loss and heat gain calculations and the Manual J calculation procedure.

Manual J is primarily a load calculation manual. However, additional information has been included in the Appendix to expand the application of the Manual J procedure and to complete the discussion of residential heating and cooling equipment selection. Information on load calculations for unique applications and on residential energy calculations can also be found in the Appendix.

This manual should not be used to estimate heating and cooling loads for commercial, industrial or institutional structures. It should not be used for residential structures which have unusual or atypical design features. Residential designs which include solariums, atriums, swimming pools and hot tubs are a few examples of atypical construction. Active or passive solar homes and underground homes are also excluded.

Introduction

The residential heating and cooling system must be selected and designed to provide comfort conditions in all occupied spaces regardless of season. Temperature, humidity, air movement and ventilation must be controlled by the system. In addition, the system must perform these functions at maximum efficiency in order to minimize energy consumption.

The load calculation is the basis for the system design. Loads **must be analyzed** if the furnace, condensing unit, fans, coils, ducts, and air terminals are to be sized correctly. Comfort, efficiency and reliability are closely related to correct sizing and selection of equipment.

A load calculation must be made for each room so that the room cooling and heating requirements can be determined. This information is needed for terminal selection, fan, and duct sizing. A load calculation must be made for the entire structure in order to properly size the heating and cooling equipment.

When equipment is oversized, efficiency is reduced, operating costs increase, and control over space conditions is lessened. Optimum efficiency and control occur when the equipment operates under full load. Since full load conditions occur only a few hours per year, properly sized equipment operates at over capacity and reduced efficiency most of the time. Oversizing the equipment aggravates this situation even more.

Slightly undersized equipment will provide comfort and efficiency most of the time, but space conditions will drift when extremes in weather occur. Overall, this is preferable to oversizing the equipment, but the increase in energy efficiency at the cost of a minor loss of comfort must be explained to the owner.

SECTION I
Heat Loss of a Structure

The design heat loss must be calculated for the winter outdoor design temperature. Because the maximum heat loss occurs during the early morning hours, before sun rise and at a time of occupant inactivity, the heat gains due to solar radiation and internal heat gains are not considered in the heat loss calculation.

When central heating equipment is installed, the equipment sizing load (design heating load) is equal to the heat loss for the entire house as shown by the "heating summary" that is located on the front of the J1 Form.

The heating load for each individual room must also be estimated to determine:

1) The room supply CFM (heating) if a "central" warm air system is installed.

2) The size of the individual room heating units (baseboard radiation for example) if a "distributed" heating system is installed.

Also refer to the most recent edition of ACCA Manual D for a complete discussion of the relationship between the Manual J heating load calculations and the residential air side (duct) design procedure.

1-1 Outside Design Condition
Outside design temperatures for many localities are listed in Table 1 (located in back of this manual). The outside design temperature is not the lowest temperature recorded. Temperature extremes occur only a few hours per year and do not represent the actual conditions experienced during average winter weather. A locality may have a record low of -20°F, but the coldest weather normally experienced might be 0°F to 5°F, and the average winter temperature might be 35°F. Equipment sized for 0°F will be over-sized for all but a few hours per year and equipment sized for -20°F will be unreasonably oversized and inefficient. The values in Table 1 are based on the 97½% values tabulated in the American Society of Heating, Refrigerating and Air Conditioning Engineers (ASHRAE) weather data. Note that in some cases the outdoor design temperature may be specified by state or local code or by a public utility.

1-2 Inside Design Conditions
The minimum indoor design temperature may be specified by local code or utility regulations. In some locations, the ASHRAE residential energy standard (Standard 90.2) may serve as the governing document. In this manual, an indoor design temperature of 70°F is recommended for heat loss calculations.

1-3 Design Temperature Difference
The heating season design temperature difference is calculated by subtracting the outdoor design temperature from the indoor design temperature.

1-4 Building Losses
Building losses are those which are associated with the building envelope such as:

A. Heat loss through glass windows and doors by conduction.

B. Heat loss through solid doors by conduction.

C. Heat loss through walls exposed to outdoor temperatures or through walls below grade.

D. Heat loss through partitions which separate spaces within the structure that are at different temperatures.

E. Heat loss through ceilings to a colder room or to an attic.

F. Heat loss through a roof-ceiling combination.

G. Heat loss through floors to a colder basement, crawl space or to the outside.

H. Heat loss through on grade slab floors or through basement floors.

I. Heat loss due to infiltration through windows and doors or through cracks and penetrations in the building envelope.

1-5 System Losses
Losses associated with operation of the heating, ventilating, cooling (HVAC) system are:

A. Heat loss through ducts located in an unheated space.

B. Ventilation air which must be heated before it is introduced into the space. (In older structures infiltration provided enough fresh air to the space making ventilation unnecessary. In newer structures, tighter construction may require ventilation).

C. Bathroom and kitchen exhaust systems tend to increase infiltration, but they are not a design factor because they are used intermittently.

D. Normally, combustion air for gas or oil fired furnaces must be provided. In older homes, infiltration will meet the combustion air requirement. In newer, tighter homes it may be necessary to introduce combustion air to the burner or the furnace room. (Refer to National Fire Protection Agency (NFPA) 54 for a complete discussion.)

1-6 Garages

If the owner requests heat in the garage, a zonal heater is preferable to heat supplied by a central system because the zonal heater can be controlled separately. In addition, air should not be returned to a central system from the garage because of the health and safety hazards that are associated with air that is contaminated by exhaust fumes. If the garage is to be heated from a central system, then outdoor air must be used to replace the air delivered to the garage. This make-up air can be provided by either infiltration or mechanical means.

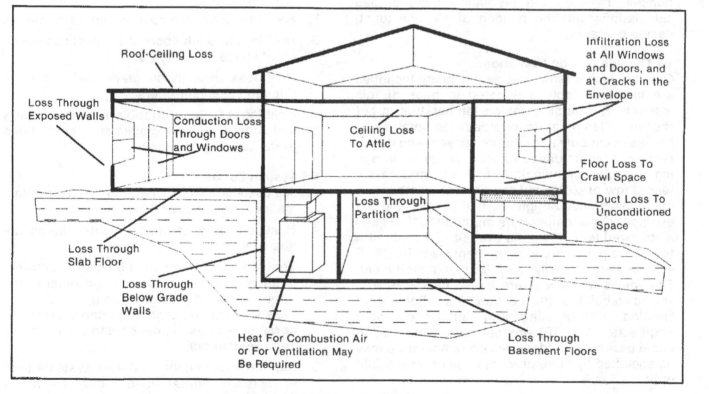

Figure 1-1 Winter Heating Loads

SECTION II
Heat Loss Calculations

Select the winter outside design temperature from Table 1 located in the back of this manual. Select the inside design temperature to meet code or owner requirements.

2-1 Window and Door Losses
Window and door designs commonly used in residential construction are illustrated in Figure 2-1.

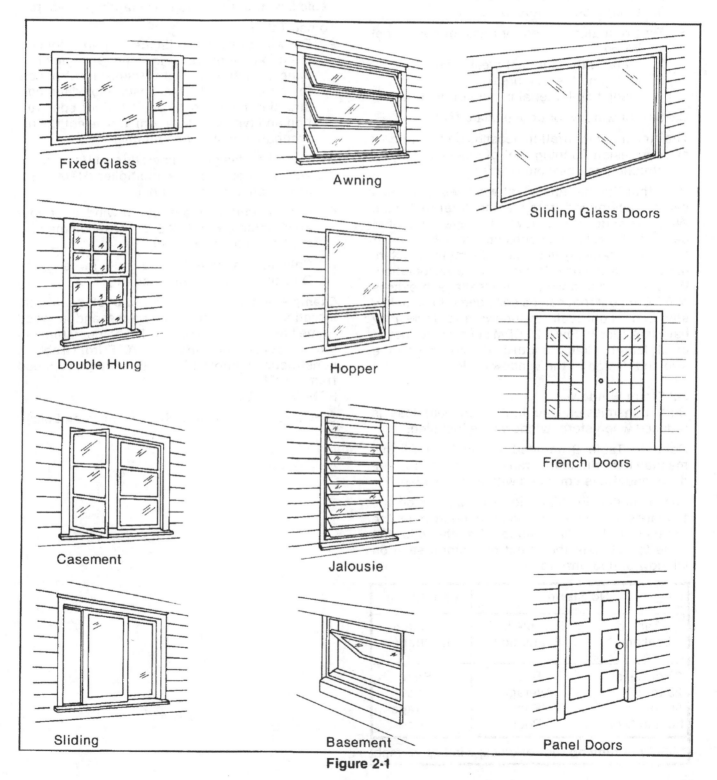

Figure 2-1

Losses associated with windows and doors are the result of heat that is transmitted through the windows or doors, and infiltration that occurs at these points. The amount of heat that is transmitted depends on the construction details of the window or door:

A. Glazing
   1. Number of panes (one, two or three)
   2. With or without storm window
   3. Emittance of glass (U value)

B. Type of material used for frame or door panel
   1. Wood
   2. Metal with a thermal break (TIM)
   3. Metal (no thermal break)
   4. Insulating Material in core of door panel

C. Size of window or door (square feet)

The amount of infiltration depends on the window or door design (running foot of crack), quality of construction and size (square feet).

Note that the "tightness" of windows and doors can be determined from manufacturer test data. Windows or doors that have a leakage of 0.5 CFM per foot of crack, when subjected to a 25 mile per hour (test chamber) wind, meet the minimum standards of the Window Manufacturers Association. Windows or doors which have leakage in excess of 0.5 CFM per foot can be considered to be below standard. High quality windows and doors may have leakages below 0.25 CFM per running foot of crack length. Crack length is determined by the size and design of the window or door.

2-2 Storm Windows

Both transmission and infiltration heat loss is reduced when storm windows are installed.

Refer to Table 2 (located at the back of this manual) to evaluate the transmission loss for windows and doors equipped with a storm sash.

The reduction in infiltration will depend on the tightness of the storm window relative to the tightness of the prime window. Use the following table to estimate the effect of a storm sash on window or door infiltration.

No Storm		With Storm
Tested Leakage *	Table 5 Evaluation	Table 5 Evaluation
.25 or Less	Best	Best
.25 to .50	Average	Best
.50 to 1.0	Poor	Average
1.0 and Over	Poor	Poor
*CFM per running foot of crack; wind @ 25 mph		

2-3 Calculation of Window and Door Losses

Window or door transmission losses can be calculated by using the heat transfer multipliers (HTM) tabulated in Table 2 located at the back of this manual. Note that Table 2 only accounts for the transmission heat loss due to the temperature difference across the glass and sash. Window and door infiltration losses are not included in the Table 2 HTM values but are considered in the Table 5 winter infiltration evaluation procedure.

To use Table 2:

A. For windows refer to the descriptions for construction numbers 1 through 9 (number of panes, emittance value, frame construction) for doors refer to the descriptions for construction numbers 10 and 11 (door construction and type of insulation) and select the appropriate construction number.

B. Use the design temperature difference to select a heat transfer multiplier (HTM) for a given construction number.

C. Determine the square feet of window or door area. Refer to Figure 2-2 for the recommended method of measurement.

D. Calculate the heat loss (Q Btu/hr) where:
   Q (Btuh) = HTM x Area (sq. ft.)

Example: A double hung window of standard construction, with single clear glass and a wood frame has an area of 9 sq. ft. The temperature difference across the window is 70°F from Table 2, construction number 1A, (single clear glass, wood frame, 70°F).

HTM = 69.3 Btuh/sq. ft.

Q = 69.3 x 9.0 = 624 Btuh (Transmission Loss)

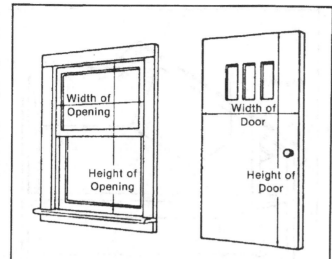

The correct way to measure individual windows or doors is illustrated above. Window or door measurements are usually recorded to the nearest 0.1 ft. The table below shows how measured inches can be converted to tenths of a foot. The total area for each window and door should be calculated by multiplying the actual width and height measurements before rounding to the nearest sq. ft.

Measured Inches	Recorded 0.1 Ft.	Measured Inches	Recorded 0.1 Ft.
1	.1	7	.6
2	.2	8	.7
3	.3	9	.8
4	.3	10	.8
5	.4	11	.9
6	.5		

Figure 2-2 Window and Door Measurement

2-4 Wall and Partition Losses Above Grade
Wall heat losses are determined from the heat that is transmitted through the wall or partition due to the temperature difference across the wall or partition. Table 2 lists heat transfer multipliers (HTM for various types of wall construction and for various design temperature differences. Losses are computed for the net wall area. (Subtract door and window areas from gross wall area).

Example: A frame wall with brick veneer, gypsum board, and with R-13 insulation in the cavity has a net area of 180 sq. ft. The room is to be maintained at 70°F when the outside design temperature is 10°F.

The design temperature difference is 60°F. From No. 12-D, Table 2, the HTM is 4.8 Btu/(hr. sq. ft).

Q = 4.8 x 180 = 864 Btuh (loss through the wall).

2-5 Above Grade Wall and Partition Areas
An exposed above grade wall is one which faces the outside. The length of exposed wall is recorded to the nearest foot. If a room has two or more exposed walls that are of the same construction, the recorded length is equal to the sum of the individual wall lengths. Wall height is recorded to the nearest one-half foot. Use an average height for walls that meet a sloped ceiling or sloped roof-ceiling. Gross exposed wall area is equal to the wall length multiplied by its height, rounded to the nearest square foot. Gross wall area includes window, and door areas. Net wall area can be computed by subtracting the window and door areas from the corresponding gross wall area. Refer to Figure 2-3 for an illustration of how above grade walls are measured.

Partitions and knee walls are walls that are not directly exposed to the outside but separate a conditioned space from an unconditioned space. Partition and knee wall areas are computed in the same manner as exposed wall areas. Refer to Figures 2-4 and 2-5 for examples of a partition wall and a knee wall.

Note that the dimensions and areas for closets and halls are usually included with those of the adjoining rooms. However, large closets and entrance halls should be considered as separate rooms. When a stair case is next to a cold outside wall the stair case wall area should be included with the exposed wall area for the hall or room below.

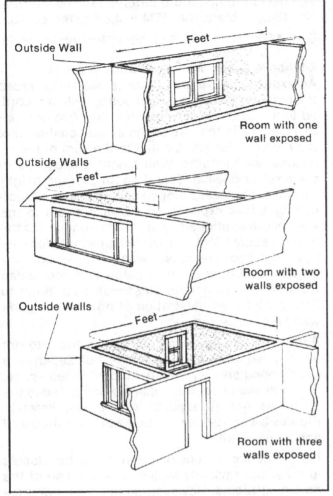

Figure 2-3 Above Grade Exposed Walls

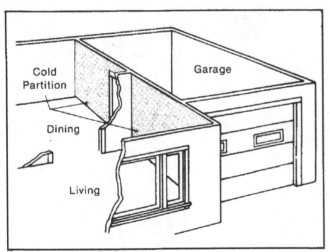

Figure 2-4 Cold Partition

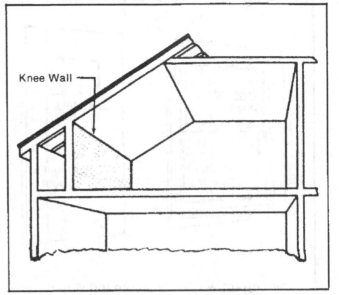

Figure 2-5 Knee Wall Partition

2-6 Wall Losses Below Grade

Temperature differences (and heat loss) across walls below grade depend on the ground temperature at various depths.

Between the surface and two feet deep, the insulating effect of the soil is slight and the ground temperature is approximately equal to winter outside design temperature. A wall that does not extend more than two feet below grade should be evaluated as an above grade wall (construction numbers 12 or 14).

At two feet or more below grade, the insulating effect of the ground is more pronounced and construction number 15 applies. When using construction number 15 the winter design temperature difference is used to select the HTM.

If the bottom of the wall extends more than two feet but not more than five feet below grade, refer to construction numbers 15 a,b,c or d for the proper HTM value. Multiply this HTM value by the net wall area that corresponds to the distance between ground level and the bottom of the wall.

If the bottom of the wall extends more than five feet below grade refer to Table 2, construction numbers 15 e,f,g or h for the proper HTM value. Multiply this HTM value by the net wall area that corresponds to the distance between ground level and the bottom of the wall.

Example: A masonry wall extends five feet below grade and is not insulated. The wall area is 620 sq. ft. The basement is 60°F and the outside design temperature is -5°F.

The design temperature difference is 65°F.

From No. 15-a, Table 2, the HTM is 8.1 Btu/(hr. sq. ft.).

Q = 8.1 Btu/(hr. sq. ft.) x 620 sq. ft. = 5,022 Btuh (loss through the wall).

2-7 Below Grade Wall Area
Walls that extend two or more feet below grade are exposed walls. Below grade wall areas are computed in the same manner as above grade wall area except that the below grade wall height is equal to the distance from the bottom of the wall to grade level. If the wall also extends above grade, the net area of the above grade strip is accounted for in a separate calculation. If the grade level varies, use an average grade level to compute the below grade wall area for each wall exposure. Refer to Figure 2-6 for an illustration of a below grade wall.

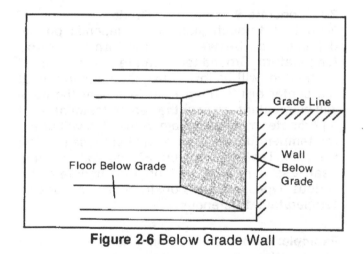

Figure 2-6 Below Grade Wall

2-8 Ceiling and Roof Losses
Heat will be lost through ceilings located beneath cold attics or unheated spaces. If the ceiling cavity is vented to the outside, assume the temperature in the cavity or attic is equal to the outside air temperature. If the cavity above a ceiling is heated, use the actual temperature difference expected. Table 2 lists the HTM for various types of ceilings, roofs, ceiling-roof combinations, and temperature differences.

Example: A ceiling with an R-19 rating is below a vented attic. The area of the ceiling is 300 sq. ft. The winter design temperature difference is 70°F.

From No. 16-D, Table 2, the HTM is 3.7 Btu/(hr. sq. ft.).

Q = 3.7 Btu/(hr. sq. ft.) x 300 (sq. ft.) = 1,110 Btuh (loss through the ceiling).

2-9 Ceiling and Roof Areas
A room has an exposed ceiling when it is located directly beneath an attic, roof or an unconditioned space. Determine the area of an exposed ceiling by multiplying room length by room width. When only part of the ceiling in a room is exposed, determine the area for the exposed portion only. Length and width are normally measured to the nearest foot and area is recorded to the nearest square foot. Refer to Figure 2-7 for an illustration of exposed ceilings.

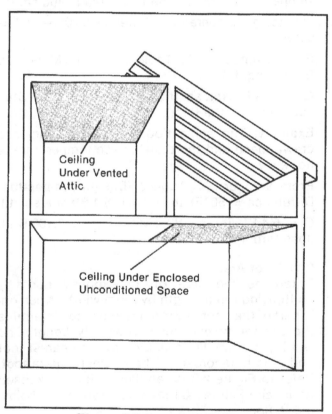

Figure 2-7 Exposed Ceiling

2-10 Floors
Table 2 lists the HTM for various types of floor construction and design temperature differences. Note that the space below the floor may be open or vented to the outside, or enclosed but not heated, or enclosed and heated to a temperature that is less than the indoor design temperature.

When computing the heat loss through the floor, use the actual temperature difference which is expected to exist across the floor on a design day and construction number 20.

If the actual temperature difference across the floor is not known, and if the space below the floor is well vented, use the design temperature difference and construction number 20.

If the actual temperature difference across the floor is not known, and if the space is enclosed and not vented (or slightly vented) use the design temperature difference and construction number 19.

Use the design temperature difference and construction number 21 for basement (below grade) floors.

Example: An uninsulated carpeted wood floor is above a well vented (open), crawl space. The room temperature is 70°F. The outside design temperature is 20°F. The floor area is 400 sq. ft.

The design temperature difference is 70° − 20° = 50°F.

From Number 20-F, Table 2, the HTM is 10.9 Btu/(hr. sq. ft.).

Q = 10.9 Btu/(hr. sq. ft.) x 400 sq. ft. = 4,360 Btuh (loss through the floor).

Example: The same floor is over an enclosed crawl space and the crawl space temperature is not known.

From Number 19-F, Table 2, (Design Temperature Difference = 50°F), the HTM is 5.4 Btu/(hr. sq. ft.).

Q = 5.4 Btu/(hr. sq. ft.) x 400 sq. ft. = 2,160 Btuh (loss through the floor).

2-11 Floor Area
Determine the area of an exposed floor by multiplying room length by room width. When only part of the floor in a room is exposed, determine the area for the exposed portion only. Length and width are normally measured to the nearest foot and area is recorded to the nearest square foot. Refer to Figure 2-8 for an illustration of exposed floors and Figure 2-6 for an illustration of a below grade floor.

Note: Do not calculate the floor area for "slab on grade" construction (refer to Section 2-12).

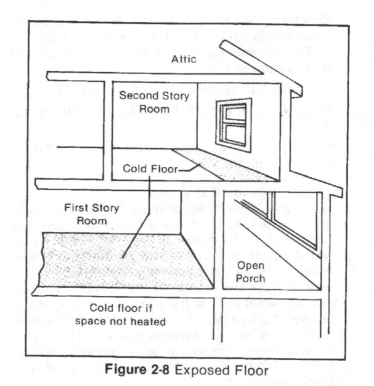

Figure 2-8 Exposed Floor

2-12 Concrete Slab Floor on Grade
Heat loss through slab floors depends on the difference between room and ground temperature. Ground temperature along the edge of the floor will be lower than the temperature at the center of the floor. Heat loss near the edge will be greater than near the center. Installation of a perimeter heating system in the slab will affect the temperature difference and heat loss through the slab. Table 2, Numbers 22 and 23, tabulates heat loss per linear foot of floor perimeter for various types of slab construction, and design temperature differences.

Example: A concrete slab on grade has one inch of insulation along the outside edge. The winter design temperature is 0°F and the room is 65°F. The slab area is 30 ft. x 20 ft.

The design temperature difference (65-0) = 65°F.

From Number 22-B, Table 2, the heat loss per foot of perimeter is 26.6 Btu/(hr. sq. ft.).

The perimeter length is: 30 + 30 + 20 + 20 = 100 ft.

Q = 26.6 Btu/(hr. ft.) x 100 ft. = 2,660 Btuh (loss through slab).

2-13 Slab on Grade Measurement

Most of the heat loss for a concrete slab on grade occurs at the outer edge so that heat loss is based on the running feet of exposed edge. This is measured and recorded to the nearest foot. Refer to Figure 1-1 for an example of slab on grade construction.

2-14 Heat Loss Due to Infiltration

The heat loss associated with winter infiltration can be calculated by using the following equation.

$$Q = 1.1 \times CFM \times (RAT\text{-}OAT)$$

Where:

Q = The heating load in Btuh
1.1 = A constant
CFM = cu.ft./min. of outdoor air infiltration
RAT = Indoor design temperature °F
OAT = The winter design temperature °F

Before the infiltration CFM can be determined, Table 5 must be used to estimate the number of heating season air changes for the entire house. After this estimate is made, calculation procedure "A" can be used to estimate the total winter infiltration CFM and to calculate a value for the winter infiltration HTM. The winter infiltration HTM and Line 12 of the J1 Form can be used to calculate the portion of the infiltration heating load which would be assessed to each room. (Manual J prorates the total infiltration to each room in proportion to the amount of window and door area associated with that room.)

Refer to Appendix A-5 for a more comprehensive air change calculation procedure if the structure is not reasonably similar to the description implied by Table 5 and the notes that go with Table 5.

2-15 Ventilation Air

Ventilation is defined as outdoor air that is mechanically introduced into the conditioned space through the heating equipment. During the heating season, mechanical ventilation may be required to dilute odors, to avoid excessive indoor humidity, to provide make up air for appliance exhaust fans and kitchen or toilet exhaust fans or to provide combustion air for a fossil fuel furnace. Refer to Table 9 (located at the back of this manual) to determine if ventilation may be required.

If mechanical ventilation is used, the heat required to temper the outdoor air can be calculated as follows:

$$Q = 1.1 \times CFM \times (RAT\text{-}OAT).$$

Where:

Q = Heat to temper the air (Btuh)
1.1 = A constant
CFM = cu.ft./min. of ventilation
RAT = Indoor design temperature °F
OAT = The winter design temperature °F

The ventilation heating load should be added to the heating load that is calculated on Line 15 of the J1 Form for the "entire house". Use the space provided on the front of the J1 Form (titled "Heating Summary") for this calculation. This combined load is the equipment load.

2-16 Duct Heat Loss

Heat loss from ducts depends on duct CFM, duct size and shape, insulation, velocity, length of the run, tightness of the duct construction, and temperature difference across duct wall. Since none of these are known until after the system has been designed, the following procedure is recommended:

A. Ignore heat loss for ducts located within the conditioned space. (This loss is a heat gain for the occupied space.)

B. Use Table 7-A to estimate duct losses for ducts which run in unheated spaces.

TABLE 7A - Duct Loss Multipliers

Case I - Supply Air Temperatures Below 120°F	Duct Loss Multipliers	
Duct Location and Insulation Value	Winter Design Below 15 °F	Winter Design Above 15°F
Exposed to Outdoor Ambient		
Attic, Garage, Exterior Wall, Open Crawl Space - None	.30	.25
Attic, Garage, Exterior Wall, Open Crawl Space - R2	.20	.15
Attic, Garage, Exterior Wall, Open Crawl Space - R4	.15	.10
Attic, Garage, Exterior Wall, Open Crawl Space - R6	.10	.05
Enclosed In Unheated Space		
Vented or Unvented Crawl Space or Basement - None	.20	.15
Vented or Unvented Crawl Space or Basement - R2	.15	.10
Vented or Unvented Crawl Space or Basement - R4	.10	.05
Vented or Unvented Crawl Space or Basement - R6	.05	.00
Duct Buried In or Under Concrete Slab		
No Edge Insulation	.25	.20
Edge Insulation R Value = 3 to 4	.15	.10
Edge Insulation R Value = 5 to 7	.10	.05
Edge Insulation R Value = 7 to 9	.05	.00
Case II - Supply Air Temperatures Above 120°F	Winter Design Below 15°F	Winter Design Above 15°F
Duct Location and Insulation Value		
Exposed to Outdoor Ambient		
Attic, Garage, Exterior Wall, Open Crawl Space - None	.35	.30
Attic, Garage, Exterior Wall, Open Crawl Space - R2	.25	.20
Attic, Garage, Exterior Wall, Open Crawl Space - R4	.20	.15
Attic, Garage, Exterior Wall, Open Crawl Space - R6	.15	.10
Enclosed In Unheated Space		
Vented or Unvented Crawl Space or Basement - None	.25	.20
Vented or Unvented Crawl Space or Basement - R2	.20	.15
Vented or Unvented Crawl Space or Basement - R4	.15	.10
Vented or Unvented Crawl Space or Basement - R6	.10	.05
Duct Buried In or Under Concrete Slab		
No Edge Insulation	.30	.25
Edge Insulation R Value = 3 to 4	.20	.15
Edge Insulation R Value = 5 to 7	.15	.10
Edge Insulation R Value = 7 to 9	.10	.05

Example: A room has a heat loss of 4,300 Btuh. The duct serving the room is located in the attic and is insulated with R=6 insulation. The room heating requirement must be increased to offset the heat loss from the duct. Assume 105°F supply temperature and winter design temperature of 5°F.

Duct loss = 0.10 x 4,300 Btuh = 430 Btuh.

Total room load = 4,300 + 430 = 4,730 Btuh.

2-17 Duct Insulation

State or local energy codes or utility standards may specify duct insulation requirements or they may refer to ASHRAE Standard 90.2.; otherwise use the following information to determine the duct insulation requirements.

Duct insulation is not required for:

• Ducts which are located in a heated space.

• Ducts which are installed in interior walls.

• "Heating only" ducts which are located in insulated basements.

• "Heating only" ducts which are located in unventilated crawl spaces and which have insulated walls.

• Ducts which are installed in exterior walls and which have a minimum of R6 wall insulation between the duct and the exterior wall surface.

Use the following guide to determine the amount of insulation that should be installed on ducts which do require insulation.

Duct Location	Recommended Level	Minimum Level
Totally Exposed to Outdoors	R6	R4
Attics and Garages	R6	R4
Open or Uninsulated Crawl Space	R6	R4
Exterior Walls, No Wall Insulation	R6	R4
Enclosed, Insulated Crawl Space	R4	R2
Uninsulated Unheated Basement	R4	R2
In or Under Concrete Slab	**	**
**Refer to ACCA Manual 4		

All ducts should have their seams sealed with tape.

2-18 Heating Equipment Capacity

The output of the heating equipment should be based on the design heating load. Electric resistance heating equipment should be sized to provide 100 percent of the design heating load requirement plus a small margin, liberal over sizing is not recommended. Fossil fuel furnaces should be sized to provide 100 percent of the design heating load requirement plus any margin that occurs because of product line capacity increments. Also refer to Section 7-26.

"Heating only" heat pumps (air source or water source) can be sized to satisfy the design heating load or these units can be slightly undersized if auxiliary heat is used to supplement the heat pump when the outdoor temperatures drop near the winter design temperature. Also refer to Section 7-28.

Heating and cooling heat pumps (air source and water source) should not be sized to satisfy the design heating load if this practice results in oversized cooling equipment which provides poor cooling comfort or performance. Section 7-27 of this manual discusses the limitations on over sizing cooling equipment. Also refer to Section 7-28.

Multi-zone heat pump equipment sizing requires a separate heating load calculation for each zone. Refer to Section 7-28 and Appendix 2 for more information on calculating residential zone loads and sizing multi-zone equipment.

Electrical resistance auxiliary heat should be sized to make up the difference between the design heating load and the heat pump heating capacity on a design day. Auxiliary heating elements should not be sized to provide emergency or stand-by heat. For more information refer to Section 7-28.

Emergency or stand-by (electrical resistance) heat should be sized to satisfy local codes or utility regulations. When emergency heat is installed at the owner's or contractor's discretion, it may be sized to provide some reasonable fraction of the design heating load. In all cases, the emergency heat should be controlled to operate independently of the primary heating system. (Refer to Section 7-28.)

The output heating capacity of furnaces or boilers used in conjunction with air source or water source heat pumps (fossil fuel assisted heat pumps) should be sized to provide 100 percent of the design heating load requirement plus any margin that occurs because of product line capacity increments.

SECTION III
Example Problem - Heat Loss Calculation

Data must be obtained from drawings, or by a field inspection, before load calculations can be made. The data required includes:

A. Measurements to determine areas:
1. Overall area of windows and doors
2. Gross areas of walls exposed to outside conditions
3. Gross area of partitions
4. Gross areas of walls below grade
5. Area of ceilings or floors adjacent to unconditioned space
6. Floor area for each room

Closets and halls are usually included with adjoining rooms. Large closets or entrance halls should be considered separately. Wall, floor, or ceiling dimensions can be rounded to the nearest foot. Window dimensions are recorded to the nearest inch. Measure the size of window or door opening. Do not include frame.

B. Construction details:
1. Window type and construction
2. Door type and construction
3. Wall construction
4. Ceiling construction
5. Roof construction
6. Floor construction

C. Temperature differences:
1. Temperature differences across components exposed to outside conditions
2. Temperature differences across partitions, floors, and ceilings adjacent to unoccupied spaces

Table 1, located in back of this manual, lists the outside design temperatures for various locations. Room temperatures are determined by the owner or builder based on recommendations by the heating and cooling contractor, or prescribed by applicable codes. ACCA recommends 70°F.

3-1 Calculation Procedure
Once areas, construction details, and temperature differences are determined, the tables in the back of this manual and the Manual J worksheet can be used to calculate heat loss. The total heat loss for a room is the sum of the heat lost through each structural component of the room. The heat lost through any component is calculated by multiplying the HTM found in the tables by the area of the component.

HTM values for temperature differences that fall between those listed in the tables can be interpolated as follows:

For example, a brick wall above grade, construction Number 12-F is subjected to a design temperature difference of 63°F. Table 2 indicates a HTM of 4.2 Btu/(hr. sq. ft.) at 60°F and 4.6 Btuh/(hr. sq. ft.) at 65°F. Select HTM = 4.4 Btu/(hr. sq. ft.) which is approximately equal to 4.2 + 3/5 (4.6-4.2) Btu/(hr. sq. ft.)

3-2 Example Problem
Figures 3-1 and 3-2 represent a house located in Cedar Rapids, Iowa. Figure 3-3 lists the construction details. Assume the inside design temperature is 70°F. From Table 1, the outside design temperature is -5°F.

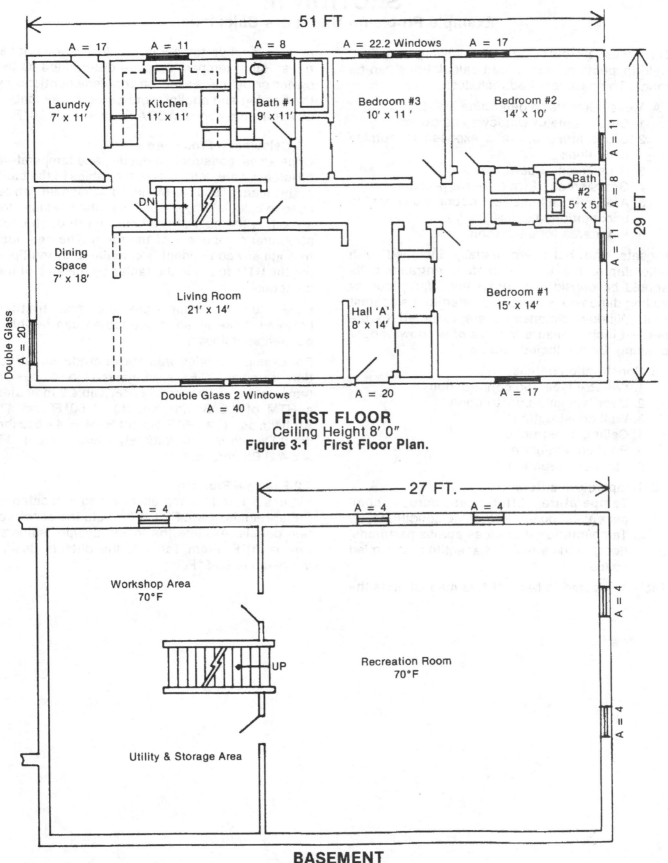

FIRST FLOOR
Ceiling Height 8' 0"
Figure 3-1   First Floor Plan.

BASEMENT
Ceiling Height Including Joist Space 8' 0"
Wall: Above Grade 3' 0"
Below Grade 5' 0"
Figure 3-2   Basement Plan.

FIGURE 3-3 EXAMPLE HEAT LOSS CALCULATION
DO NOT WRITE IN SHADED BLOCKS

				Entire House			1 Living			2 Dining			3 Laundry			4 Kitchen			5 Bath-1		
1 Name of Room				Entire House			1 Living			2 Dining			3 Laundry			4 Kitchen			5 Bath-1		
2 Running Ft. Exposed Wall				160			21			25			18			11			9		
3 Room Dimensions Ft.				51 x 29			21 x 14			7 x 18			7 x 11			11 x 11			9 x 11		
4 Ceiling Ht. Ft Directions Room Faces				8			8 West			8 North			8			8 East			8 East		

TYPE OF EXPOSURE		Const. No.	HTM Htg.	HTM Clg.	Area or Length	Btuh Htg.	Btuh Clg.	Area or Length	Btuh Htg.	Btuh Clg.	Area or Length	Btuh Htg.	Btuh Clg.	Area or Length	Btuh Htg.	Btuh Clg.	Area or Length	Btuh Htg.	Btuh Clg.	Area or Length	Btuh Htg.	Btuh Clg.
5 Gross	a	12-d			1280			168			200			144			88			72		
Exposed	b	14-b			480																	
Walls &	c	15-b			800																	
Partitions	d																					
6 Windows	a	3-A	41.3		60	2478		40	1652		20	826										
& Glass	b	2-C	48.8		20	976																
Doors Htg.	c	2-A	35.6		105	3738											11	392		8	285	
	d																					
7 Windows		North																				
& Glass		E&W																				
Doors Clg.		South																				
8 Other Doors		11-E	14.3		37	529								17	243							
9 Net	a	12-d	6.0		1078	6468		128	768		180	1080		127	762		77	462		64	384	
Exposed	b	14-b	10.8		460	4968																
Walls &	c	15-b	5.5		800	4400																
Partitions	d																					
10 Ceilings	a	16-d	4.0		1479	5916		294	1176		126	504		77	308		121	484		99	396	
	b																					
11 Floors	a	21-a	1.8		1479	2662																
	b																					
12 Infiltration HTM			70.6		222	15673		40	2824		20	1412		17	1200		11	777		8	565	
13 Sub Total Btuh Loss =6+8+9+10+11+12						47808			6420			3822			2513			2115			1630	
14 Duct Btuh Loss			0%		—			—			—			—			—					
15 Total Btuh Loss = 13+14						47808			6420			3822			2513			2115			1630	
16 People @ 300 & Appliances 1200																						
17 Sensible Btuh Gain =7+8+9+10+11+12+16																						
18 Duct Btuh Gain				%																		
19 Total Sensible Gain = 17+18																						

From Table 2

ASSUMED DESIGN CONDITIONS AND CONSTRUCTION (Heating):

		Const. No.	HTM
A.	Determing Outside Design Temperature -5° db-Table 1		
B.	Select Inside Design Temperature 70°db		
C.	Design Temperature Difference: 75 Degrees		
D.	Windows: Living Room & Dining Room - Clear Fixed Glass, Double Glazed - Wood Frame - Table 2 .	3A	41.3
	Basement - Clear Glass Metal Casement Windows, with Storm - Table 2	2C	48.8
	Others - Double Hung, Clear, Single Glass and Storm, Wood Frame - Table 2	2A	35.6
E.	Doors: Metal, Urethane Core, no Storm - Table 2	11E	14.3
F.	First Floor Walls: Basic Frame Construction with ½" Asphalt Board (R-11) - Table 2	12d	6.0
	Basement wall: 8" Concrete Block - Table 2		
	Above Grade Height: 3 ft (R = 5)	14b	10.8
	Below Grade Height: 5 ft (R = 5)	15b	5.5
G.	Ceiling: Basic Construction Under Vented Attic with Insulation (R-19) - Table 2	16d	4.0
H.	Floor: Basement Floor, 4" Concrete - Table 2	21a	1.8
I.	All moveable windows and doors have certified leakage of 0.5 CFM per running foot of crack (without storm), envelope has plastic vapor barrier and major cracks and penetrations have been sealed with caulking material, no fireplace, all exhausts and vents are dampered, all ducts taped.		

DO NOT WRITE IN SHADED BLOCKS

6 Bedroom 3			7 Bedroom 2			8 Bath 2			9 Bedroom 1			10 Hall			11 Rec. Room			12 Shop & Utility			
10			24			5			29			8			83			77				2
10 x 11			14 x 10			5 x 5			15 x 14			8 x 14			27 x 29			24 x 29				3
8	East		8	E & S		8	South		8	S & W		8	West		8	E & S		8	East			4
Area or Length	Btuh Htg	Clg	Area or Length	Btuh Htg	Clg	Area or Length	Btuh Htg	Clg	Area or Length	Btuh Htg	Clg	Area or Length	Btuh Htg	Clg	Area or Length	Btuh Htg	Clg	Area or Length	Btuh Htg	Clg	
80			192			40			232			64										5
															249			231			
															415			385			
															16	781		4	195		6
22	783		28	997		8	285		28	997												
																					7
												20	286								8
58	348		164	984		32	192		204	1224		44	264								9
															233	2516		227	2452		
															415	2283		385	2118		
110	440		140	560		25	100		210	840		112	448								10
															783	1409		696	1253		11
22	1553		28	1977		8	565		28	1977		20	1412		16	1130		4	282		12
	3124			4518			1142			5038			2410			8119			6300		13
	—			—			—			—			—			—			—		14
	3124			4518			1142			5038			2410			8119			6300		15
																					16
																					17
																					18
																					19

Figure 3-3 illustrates the completed worksheet. Here is a line by line explanation of the procedure.

Line 1. Identify each area that is heated.

Lines 2 and 3. Enter room dimensions. The dimensions shown were from Figures 3-1 and 3-2.

Line 4. Enter the ceiling height for reference. The direction the room faces is not a concern when making the heat loss calculation, but is used in the heat gain calculations.

Line 5A through 5D. Enter gross area for walls. For rooms with more than one exposure, use one line for each exposure. For rooms with more than one type of wall construction, use one line for each type of construction. Find the construction number in the tables in the back of this manual. Enter this number on the appropriate line.

Example: The gross area of west living room wall is 168 sq. ft. This wall is listed in Table 2, the construction number is 12-D.

Lines 6A through 6C. Enter the area and orientation of windows and glass doors for each room. Determine construction numbers from the tables and enter them. Determine temperature difference across the glass and read HTM for heating from the tables. Enter these. Multiply the window area by its HTM to determine the heat loss through that window. Enter the heat loss in the column marked Btuh heating.

24

Example: The living room has 40 sq. ft. of wood frame fixed double glass windows. Construction number of the window is 3A. The temperature difference across the window will be based on winter design conditions. The design temperature difference is 70° -(-5°) = 75°F. The HTM listed for 75°F on line 1-C of the tables is 41.3 Btu/(hr. sq. ft.). The heat loss through the window is 40 sq. ft. x 41.3 Btu/(hr. sq. ft.) = 1,652.

Example: The workshop has 4 sq. ft. of metal frame awning glass (plus storm) windows. The construction number is 2C. If the shop temperature is 70°F, the design temperature difference is 70 - (-5) = 75°F. The HTM is 48.8 Btuh/(hr. sq. ft.) Heat loss through the window is 4 sq. ft. x 48.8 Btu/(hr. sq. ft.) = 195 Btuh.

Line 7. Not required for the heating calculation.

Line 8. Enter the area, construction number, and HTM for wood or metal doors. Multiply HTM by the door area and enter the heat loss through the door.

Example: The main entrance (hall A) has a 20 sq. ft. metal-urethane core door. The construction number of the door is 11-E. The design temperature difference is 70° - (-5) = 75°F. The HTM is 14.3 Btu/(hr. sq. ft.). The heat loss through the door is 14.3 Btu/(hr. sq. ft.) x 20 sq. ft. = 286 Btuh.

Lines 9A through 9D. For each room, subtract window and door areas from corresponding gross wall and enter net wall areas and corresponding construction number. Determine temperature difference across each wall and enter the HTM. Multiply the HTM by the wall area and enter heat loss through the wall.

Example: The west wall in the living room has a net area of 128 sq. ft. (168 sq. ft. - 40 sq. ft.) The wall construction number is 12-D. The temperature difference is 75°F and the HTM is 6.0 Btu/(hr. sq. ft.). The heat loss through the wall is 128 sq. ft. x 6.0 Btu/(hr. sq. ft.) = 768 Btuh.

Example: The basement wall surrounding the recreation room has a net area of 233 sq. ft. (above grade) and the construction number is 14-B. The HTM for a 75°F temperature difference is 10.8 Btu/(hr. sq. ft.). Heat loss above grade is 10.8 x 233 = 2,516 Btuh. Net area below grade is 415 sq. ft. The construction number is 15-B. The HTM is 5.5 Btu/(hr. sq. ft.). The heat loss below grade is 415 sq. ft. x 5.5 Btu/(hr. sq. ft.) = 2,283 Btuh.

Lines 10A and 10B. Enter ceiling area and construction number for ceilings exposed to a temperature difference. Determine the temperature difference across the ceiling and enter the HTM. Multiply the HTM by the ceiling area and enter heat loss through ceiling.

Example: R-19 insulated living room ceiling has an area of 294 sq. ft. The construction number is 16-D. Since the attic is vented, the temperature difference is (70° - (-5°)) = 75°F. The HTM is 4.0 Btu/(hr. sq. ft.) and heat loss through the ceiling is 294 sq. ft. x 4.0 Btu/(hr. sq. ft.) = 1,176 Btuh.

Example: The recreation room ceiling will have no heat loss since the temperature difference is zero.

Lines 11A and 11B. Enter floor area and floor construction number for floors subject to a temperature difference. Determine the HTM and calculate heat loss through the floor.

Example: The recreation room slab floor has an area of 783 sq. ft. The construction number is 21. The design temperature difference is 75°F and the HTM is 1.8 Btu/(hr. sq. ft.) Heat loss through the floor is 783 sq. ft. x 1.8 Btu/(hr. sq. ft.) = 1,409 Btuh.

Line 12. Use Table 5 and Calculation Procedure A to calculate the winter infiltration HTM, and enter this value on Line 12. For each room, enter the total sq. ft. of the window and door openings on Line 12. Finally, compute the infiltration heat loss due to infiltration for each room by multiplying the winter infiltration HTM value by the appropriate sq. ft. of window and door opening. Enter these results on Line 12. Refer to Figure 3-4 for an example of the infiltration HTM calculation.

Line 13. Calculate the subtotal heat loss for each room and for the entire house. (Add lines 6,8,9,10,11,12.)

Line 14. Calculate and enter duct heat loss for each room. In this problem, the ducts are in the heated space so duct losses can be ignored. For duct loss calculation procedure, see paragraph 2-16.

Line 15. Add the duct losses to the room losses. This sum is the total heat required for each room and for the structure. If ventilation air is not introduced through the equipment, the sum for the "entire house" column can be used to size the heating equipment.

If ventilation air is used, the heat required to temper this air must be added to the total heat required by the structure. This calculation can be made on the front of the Manual J Form in the panel titled "Heating Summary."

In either case, the output capacity of the heating equipment shall not be less than the calculated loss.

Since ventilation was not included in the example problem, the design heating load is calculated as follows:

Line 15 Heat Loss (Btuh) = 47,808 (Entire House)
Ventilation CFM = 0
Ventilation Heat (Btuh) = 0
Design Heating Load = 47,808 (House)
                       + 0 (Vent)
                       = 47,808 Btuh

If this problem had included 100 CFM of outside air for ventilation, the design heating load would be calculated as follows:

Line 15 Heat Loss (Btuh) = 47,808 (Entire House)
Ventilation CFM = 100
Design Temperature Difference = 75 °F
Ventilation Heat (Btuh) = 1.1 x 100 CFM x 75 °F = 8,250 Btuh
Design heating Load = 47,808 + 8,250 = 56,058 Btuh.

Table 5

Infiltration Evaluation

Winter air changes per hour

Floor Area	900 or less	900 - 1500	1500 - 2100	over 2100
Best	0.4	0.4	0.3	0.3
Average	1.2	1.0	0.8	0.7*
Poor	2.2	1.6	1.2	1.0

For each fire place add:		Best	Average	Poor
		0.1	0.2	0.6

Average - Plastic vapor barrier, major cracks and penetrations sealed, tested leakage of windows and doors between 0.25 and 0.50 CFM per running foot of crack, electrical fixtures which penetrate the envelope not taped or gasketed, vents and exhaust fans dampered, combustion air from indoors, intermittent ignition and flue damper, some duct leakage to unconditioned space.

Procedure A - Winter Infiltration HTM Calculation

1. Winter Infiltration CFM
   0.70 AC/HR x 16269 Cu. Ft. x 0.0167   = 190 CFM
               Volume

2. Winter Infiltration Btuh
   1.1 x 190 CFM x 75 Winter TD   = 15675 Btuh

3. Winter Infiltration HTM
   15675 Btuh ÷ 222 Total Window   = 70.6 HTM
            & Door Area

* Includes Full Basement

Above Grade Volume = 51x29x(8 + 3) = 16269

Figure 3-4 Infiltration HTM Calculation

SECTION IV
Heat Gain of a Structure

The heat gain for a structure must be calculated for the summer design condition which must consider:

A. The summer outside design temperature.
B. Radiation from the sun.
C. Heat and moisture given off by equipment and appliances.
D. Heat and moisture given off by people.
E. Heat and moisture gained by infiltration.

These conditions may or may not occur simultaneously.

Example: A kitchen/dining room with west facing glass, would have a maximum heat gain between 4 and 6 p.m. in July or August, with all the loads occurring simultaneously. The outside air temperature would be high: solar radiation would be at peak intensity; the appliances would be in operation; and people would be in the room.

A bedroom with east facing windows would not experience all of the loads simultaneously. The outside air temperature would reach its peak late in the afternoon; the solar radiation would peak in the morning; appliances are not normally found in the bedroom and the bedroom is usually unoccupied during the day.

4-1 Outside Design Condition
The summer design conditions listed in Table 1 are based on the ASHRAE 2.5 percent weather data. The design temperatures are values which will be exceeded between 50 and 100 hours during the summer months (June through September). This is acceptable since equipment selected for the worst possible case would be oversized for all but a few hours when the extreme conditions occur. The following reasoning applies:

A. Normal summer temperatures impose a smaller cooling load on the structure.
B. Equipment capacity is rated at 95°F outside temperature by the Air-Conditioning and Refrigeration Institute (ARI). When the outside air temperature is below 95°F, the capacity of the equipment increases.
C. Equipment sized for extreme conditions will operate over 97.5% of the possible load hours at reduced load and increased capacity.

4-2 Daily Range
Outside temperatures usually reach their highest levels late in the afternoon and drop to their lowest levels just before daylight. The difference between the normal high and low temperatures is the daily range. Daily range is significant because the daytime cooling load is reduced when the structure is cooled by low nighttime temperatures. The daily temperature range is described as low when the difference is less than 15°F, medium when the difference is 15° to 25°F, and high when the difference is greater than 25°F.

4-3 Inside Design Condition
Use 75°F and 50% to 55% relative humidity for the inside design condition unless the owner, builder or codes specify otherwise.

4-4 Design Temperature Difference
The cooling season design temperature difference is calculated by subtracting the indoor design temperature from the outdoor design temperature.

4-5 Sun Position
The intensity with which the sun strikes a surface varies hour by hour during the day and month by month during the year. East exposures have peak intensity in the morning in July or August. West exposures and roofs peak late in the afternoon in July or August. South exposures peak in early afternoon in September or October. North exposures are not significantly affected by the sun. Since residential equipment normally does not have the ability to control rooms on an individual basis, inside temperatures can be expected to drift as the sun moves across the sky. The procedures outlined should limit room temperature swings to approximately 3°F.

4-6 Storage
Solar radiation entering through the windows will not produce a cooling load on the equipment immediately. The radiation is converted to heat when it strikes an interior surface. This heat is initially stored as the structure warms up. At a later time, the warmer surfaces begin to transfer heat to the air in the room. The room air carries this heat to the cooling coil and the cooling load on the equipment increases.

When solar radiation strikes an exterior wall or roof, the surface is heated and the heat gradually penetrates the wall or roof. At a later time, the inside surface temperature increases and heat is transferred to the air in the room. The room air carries the heat to the coil where it appears as a cooling load. The net effect of storage is to delay and smooth out the solar loads. The procedures outlined in this manual take into account the effects of storage.

4-7 Building Gains

Building heat gains are associated with heat transferred through the building envelope and with the internal loads produced by lights, people, and equipment such as:

A. Heat gained by solar radiation through glass.
B. Heat transmitted through glass by conduction.
C. Heat transmitted through walls exposed to outside air.
D. Heat gained through partitions which separate conditioned and unconditioned spaces.
E. Heat gained through ceilings from the attic.
F. Heat gained through roofs and roof/ceiling combinations.
G. Heat gained through ceilings and floors that separate conditioned and unconditioned spaces.
H. Heat gained by infiltration through doors and windows or building envelope.
I. Heat produced by people.
J. Heat produced by lights.
K. Heat produced by appliances and equipment.

4-8 Systems Gains

System gains are those associated with operation of the HVAC system such as:

A. Heat gained through ducts located in unconditioned spaces.
B. Heat gains associated with ventilation air mechanically introduced through the system.
C Bathroom and kitchen exhaust tend to increase infiltration. Note, these devices only operate a small part of the time.

4-9 Walls Below Grade and Slabs

Heat transfer through walls below grade and slabs on grade is not a factor in cooling load calculation.

4-10 Latent Loads

Moisture entering the conditioned space can be attributed to people and appliances, moisture migrating through the building envelope, and infiltration. The magnitude of the latent loads will depend on the building construction, the habits of the occupants, and the geographical location of the structure.

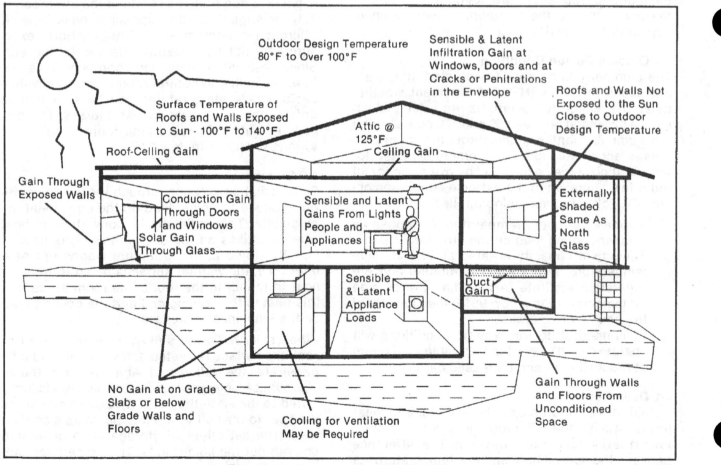

Figure 4-1 Summer Cooling Loads

SECTION V
Heat Gain Calculations

The load calculation should be based on the combined effects of transmission, convection, radiation, infiltration, and internal loads. Since these load components vary with the time of day and by the month, it may not be obvious when they combine to produce a maximum load. Indeed, the most difficult part of making a load calculation is choosing the month and time of day when the load components combine to produce the design load. The simplified 24 hour load calculation procedures for residential design loads outlined in this manual are based on the following assumptions:

A. That the amount of glass used in the structure is approximately 10 percent to 25 percent of the floor area. Special architectural treatments such as glass walls, solariums, atriums or liberal use of glass require a more detailed calculation procedure.

B. That solar gains and transmission loads for any room must be based on average values as opposed to the peak load that a room can experience. Studies of residential loads show that the cooling equipment will be considerably oversized if all the rooms are designed for their peak loads. (Common sense indicates that it is not possible for all the rooms to peak simultaneously.) The use of average values will result in a cooling unit that is sized to meet the load for the entire structure. A room temperature swing of approximately 3°F is assumed when average values are used.

   1. Transmission through walls and roofs is calculated by using average equivalent temperature difference data which is independent of the direction the wall faces.

   2. Solar gain through windows is calculated for each exposure by using average values for the exposure in question.

C. That the daily range does have an effect on the cooling load, and that the daily range will be considered either low (L), medium (M) or high (H).

D. That the system will be operated on a 24-hour per day basis and that the thermostat will be set at the indoor summer design temperature.

E. That the sensible loads associated with people, lights and minor appliances are included in the 300 Btuh allowance that is applied to each occupant.

F. That no provision has been made for unusual loads, such as entertaining groups of people.

G. That the sensible loads which are associated with common kitchen appliances are equal to an effective sensible load of 1,200 Btuh.

H. That intermittent use of kitchen and bathroom exhaust fans will not significantly affect the calculated infiltration load.

I. That the latent load on the equipment is equal to the latent load produced by ventilation, infiltration, occupants, appliances and plumbing fixtures. Refer to Section 7-22 for a complete discussion on how these latent loads are calculated.

J. That no provision has been made for unusual latent loads, such as a large number of plants or a hot tub.

K. That ventilation (when it is necessary) produces both sensible and latent loads on the equipment.

L. That the cooling equipment will be selected to satisfy both the sensible and latent loads at the summer design condition and that the manufacturer's performance data will be used to confirm that both the required sensible and latent capacity is available at this condition.

M. That structures which are zoned require a separate cooling load calculation for each zone and that this calculation includes an allowance for peak loads. (Refer to the Appendix 2 for more information on calculating residential zone loads.

5-1 Building Component Areas
Refer to Section 2 for information on measuring and calculating window, door, wall, ceiling and floor areas. Areas of walls below grade, basement floors and ground slabs are not required for the cooling load estimate.

5-2 Heat Gain Through Windows
The combined effect of radiation and transmission through vertical glass for various exposures and sky-lights (which are orientated at various angles of incidence and exposures) is given by the HTM multipliers found in Table 3. These HTM values are listed for single, double or triple glass; for glass with no shading, internal shading by external shade screens or external shading by projections and over-hangs. Multipliers for tinted glass, reflective coatings and low emittance coatings are also included. No allowance for infiltration is included in the Table 3 HTM multipliers.

It is important to identify glass that is externally

shaded since external shading produces a significant reduction in the glass HTM. For windows that are partially shaded by overhangs, use Table 8 to determine the shaded area. The heat gain through the shaded area is based on the HTM for external shading or for north-facing glass. The heat gain through the area exposed to the sun is calculated in the usual manner.

Examples:

A. An east facing window has an area of 15 sq. ft. and it has double glazing and draperies. The outside design temperature difference is 20°F. The HTM from Table 3A is 46 Btu/(hr. sq. ft.).

   $Q = 15$ sq. ft. x 46 Btu/(hr. sq. ft.) = 690 Btuh.

B. If the same window is equipped with an external shade screen that has a shading coefficient equal to 0.25, the HTM value from Table 3C is 24 Btu/(hr. sq. ft.).

   $Q = 15$ sq. ft. x 24 Btu/(hr. sq. ft.) = 360 Btuh.

C. A south facing window has an area of 22 sq. ft. and it has a single pane of heat absorbing glass with no shading. The design temperature difference is 25°F. The HTM from Table 3A is 40 Btu/(hr. sq. ft.).

   $Q = 22$ sq. ft. x 40 Btu/(hr. sq. ft.) = 880 Btuh.

D. The same south window has an 18-inch roof overhang. The window is 4 ft. x 5.5 ft. The top of the window is 8" below the overhang. The latitude is 40°F.

   (Refer to the Table 8 calculation below.) The area of shaded glass is 17.76 sq. ft. The area of glass exposed to the sun is 4.24 sq. ft.

   From Table 3, the HTM are:
   A. Window in shade = 28 Btu/(hr. sq. ft.)
   B. Window in sun = 40 Btu/(hr. sq. ft.)
   The total heat gain is:
   $Q = (28 \times 17.76) + (40 \times 4.24) = 667$ Btuh.

5-3 Heat Gain Through Walls and Partitions

Table 4 gives HTM for walls and partitions. The HTM for walls include thermal storage and the effect of the sun striking the wall. Multipliers for partitions only account for storage and the air temperature difference across the partition.

To calculate the heat gain through walls or partitions, subtract the window and door areas from the gross wall area. The net wall area is used with the HTM to determine the wall gain. Walls below grade are not a consideration because the ground temperatures are considerably lower than the outside air temperature.

Examples:

A. A wood frame wall with brick veneer has a net area of 200 sq. ft. The wall has R-13 batt insulation and is finished with plaster (gypsum) board. The summer design temperature difference is 20°F-M.

   From Table 4 (12-d), the HTM is 1.9 Btu/(hr. sq. ft.).

   $Q = 200$ sq. ft. x 1.9 Btu/(hr. sq. ft.) = 380 Btuh.

B. A wood frame partition with plaster (gypsum) board on both sides is not insulated. The net wall area is 100 sq. ft. The temperature difference across the wall is 15°F-M.

   From Table 4, (13-a), the HTM is 2.7 Btu/(hr. sq. ft.).

   $Q = 100$ sq. ft. x 2.7 Btu/(hr. sq. ft.) = 270 Btuh.

5-4 Heat Gain Through Ceilings and Floors

The HTM for roofs, ceilings, and roof/ceiling combinations are given in Table 4. The effect of solar radiation is included in the roof or roof/ceiling multipliers. The multiplier values are based on the assumption that the attic or roof/ceiling combination meets the HUD-FHA ventilation standard. For convenience, this standard is outlined by Figure 5-1 in this section.

Shade Line Multiplier
40 Degrees North Lat.

Direction Window Faces	LAT 40°
E or W	0.81
SE or SW	1.25
South	2.60

Shade Line

			Shaded Glass Area	
			Exposure	South
A	Direction Wall Faces			
B	Overhang Dimension		x in ft.	1.50
C	Shade Line Multiplier		See Table	2.60
D	Distance Between Overhang & Shade Line (z in ft.)		(BXC) =	3.90
E	Distance Between Overhang & Top of Window		y in ft.	.67
F	Shaded Height of Glass (ft)		(D-E)	3.23
G	Height of Window		w in ft.	4.00
H	Unshaded Height of Glass (ft)		(G-F) =	.77
I	Width of Window		(Feet)	5.50
J	Area, Shaded Glass (sq. ft.)		(FXI) =	17.77
K	Area, Unshaded Glass (sq. ft.)		(HXI) =	4.24

The multipliers for floors, ceilings, or floor/ceiling combinations which are located directly above or below an unconditioned space account for thermal storage and transmission due to the air temperature difference across the structure.

Floors on or below grade need not be included in the calculation.

Examples:

A. A ceiling located under a vented attic has a R-19 rating. The ceiling area is 400 sq. ft. The roof is dark. The design temperature difference is 20°F-M.

From Table 4 (16-D-dark roof), the HTM is 2.3 Btu/(hr. sq. ft.).

Q = 400 sq. ft. x 2.3 Btu/(hr. sq. ft.) = 920 Btuh.

B. A dark roof on exposed beams is constructed with 1½ inch planking and two inches of insulating board R-8). There is no ceiling. The roof area is 400 sq. ft. The design temperature difference is 20°F-M. Insulation is R-9.

From Table 4 (17-E-dark roof), the HTM is 3.9 Btu(hr. sq. ft.).

Q = 400 sq. ft. x 3.9 Btu/(hr. sq. ft.) = 1,560 Btuh.

C. A floor over an open crawl space is not insulated, but does have carpeting. The outside design temperature is 20°F higher than the room temperature. The floor area is 300 sq. ft. Temperature range is high.

From Table 4 (20-F), the HTM is 2.5 Btu/(hr. sq. ft.).

Q = 300 sq. ft. x 2.5 Btu/(hr. sq. ft.) = 750 Btuh.

5-5 Sensible Heat Gain Due to People

Add 300 Btuh for each person normally expected to occupy the house. Include this load for the rooms in which they would be found during peak load conditions. (Usually the family room or dining room). Do not include this load in rooms which are normally unoccupied when the transmission and solar loads are at their peak. Usually the number of people is estimated to be twice the number of bedrooms.

5-6 Sensible Heat Gain Due to Appliances

Include 1,200 Btuh as a sensible load in the kitchen. This is considerably less than the rated output of the appliances, but intermittent operation and exhaust hoods make this a reasonable value.

5-7 Heat Gains Due to Infiltration

Infiltration produces both sensible and latent heat gains. The sensible heat gain is calculated by using the following equation.

Q = 1.1 x CFM x (OAT-RAT)   (Sensible Gain)

Where:

Q = Sensible Cooling Load in Btuh
1.1 = A Constant Required for Consistent Units
CFM = Cubic Feet Per Minute of Infiltration
OAT = Outdoor Design Temperature °F
RAT = Indoor Design Temperature °F

Before the summer infiltration CFM can be determined, Table 5 must be used to estimate the infiltration rate in terms of air changes per hour for the entire house. After this estimate is made, calculation procedure "B" can be used to estimate the total summer infiltration CFM and to calculate a value for the summer infiltration (sensible) HTM. The summer infiltration HTM and Line 12 of the J1 Form can be used to calculate the portion of the sensible infiltration cooling load which should be assessed to each room. (Manual J prorates the total infiltration to each room in proportion to the amount of window and door area associated with that room.)

The latent infiltration gain is calculated by using the equation shown below.

Q = 0.68 x CFM x Gr   (Latent Gain)

Where:

Q = Latent cooling load in Btuh
0.68 = A constant
CFM = Cubic feet per minute of infiltration
GR = The difference in grain of moisture between the outdoor air design condition and indoor air design condition. Refer to Table 1 for a listing of the outdoor design grain difference for various cities.

Calculation Procedure "C" (on the back of the J-1 Form) should be used to calculate the latent infiltration load for the entire house. Note that the latent infiltration load for each room is not required because single zone residential equipment is not able to prorate its latent capacity to individual rooms. However, multi-zone systems which employ multiple indoor direct expansion coils are able to control latent loads on a zonal basis. In this case, the calculation Procedure "C" can be applied to each zone.

Refer to Appendix 5 for a more comprehensive air change calculation procedure if the structure is not reasonably similar to the description implied by Table 5 and the notes that go with Table 5.

5-8 Duct Gains

Duct heat gain depends on duct CFM, duct size and shape, insulation, velocity, length of the run, tightness of the duct construction, and temperature difference across duct wall. Since none of these are known until after the system has been designed, the following procedure is recommended:

A. Ignore heat gains for ducts located within the conditioned space.
B. Use Table 7-B to estimate the duct gains for ducts which run in unconditioned spaces.

Example: A room has a sensible gain of 2,100 Btuh. The duct for this room is located in the attic and is insulated with blanket insulation (R-6). The room load (and room CFM) must be increased to account for the duct loss.

Duct loss = 0.10 x 2,100 Btuh = 210 Btuh.

Total room load = 2,100 Btuh + 210 Btuh = 2,310 Btuh.

TABLE 7B · Duct Gain Multipliers

Duct Location and Insulation Value	Duct Gain Multiplier
Exposed to Outdoor Ambient	
Attic, Garage, Exterior Wall, Open Crawl Space · None	.30
Attic, Garage, Exterior Wall, Open Crawl Space · R2	.20
Attic, Garage, Exterior Wall, Open Crawl Space · R4	.15
Attic, Garage, Exterior Wall, Open Crawl Space · R6	.10
Enclosed In Unconditioned Space	
Vented or Unvented Crawl Space or Basement · None	.15
Vented or Unvented Crawl Space or Basement · R2	.10
Vented or Unvented Crawl Space or Basement · R4	.05
Vented or Unvented Crawl Space or Basement · R6	.00
Duct Buried In or Under Concrete Slab	
No Edge Insulation	.10
Edge Insulation R Value = 3 to 4	.05
Edge Insulation R Value = 5 to 7	.00
Edge Insulation R Value = 7 to 9	.00
This Table can also be found on Page 87 at the back of this manual.	

5-9 Duct Insulation

State or local energy codes or utility standards may specify duct insulation requirements or they may refer to ASHRAE Standard 90.2; otherwise refer to Section 2-11 of this manual to determine when duct insulation is required for heating. Note that the duct insulation requirements for cooling are similar except that duct insulation is required for:

• Cooling ducts which are located in unconditioned (insulated or uninsulated) basements.
• Cooling ducts located in any crawl space (whether vented or unvented, insulated or uninsulated).

Use the following guide to determine the amount of insulation that should be installed on ducts which do require insulation.

Duct Location	Recommended Level	Minimum Level
Totally Exposed to Outdoors	R6	R4
Attics and Garages	R6	R4
Open or Uninsulated Crawl Space	R6	R4
Exterior Walls, No Wall Insulation	R6	R4
Enclosed, Insulated Crawl Space	R4	R2
Unconditioned Basement	R4	R2
In or Under Concrete Slab	**	**
**Refer to ACCA Manual 4		

A suitable vapor barrier should be installed over all external duct insulation. All ducts should have their seams sealed with tape.

In this manual, a value of 230 Btuh per person is used to estimate the latent loads that are associated with occupants, appliances and plumbing fixtures. No allowance for an unusual number of plants is included in this figure. Use calculation Procedure D (on the back of the J-1 Form) to estimate the latent loads that are produced by these internal sources.

Refer to Sections 5-7 and 5-11 of this manual for a discussion of the calculations and procedures that are used to estimate the size of the latent loads produced by infiltration and ventilation.

Use calculation Procedure "D" to estimate the total latent load on the equipment.

5-11 Ventilation

Ventilation is defined as outdoor air that is mechanically introduced into the conditioned space through the heating and cooling equipment. Mechanical ventilation may be required to introduce "fresh" outdoor air into an unusually tight house to dilute odors and to control possible toxic gas buildup. In some cases, ventilation may be required to provide make up air for appliance exhaust fans, and kitchen or toilet exhaust fans. Refer to Table 9 (located at the back of this manual) for a procedure which will help determine if ventilation is required.

If mechanical ventilation is used, the sensible and latent loads associated with the ventilation CFM become a load on the cooling equipment and should be calculated as follows:

$$Q \text{ (Sense)} = 1.1 \times CFM \times (OAT\text{-}RAT)$$
$$Q \text{ (Lat)} = 0.68 \times CFM \times Gr$$

Where:
Q = Sensible or Latent Load in Btuh
    1.1 and 0.68 are constants
CFM = Cubic Feet Per Minute of Ventilation
RAT = Room Air Temperature °F
OAT = Outside Design Air Temperature °F
Gr = The difference in grains of moisture between the outdoor air design condition and indoor air design condition.

Use calculation Procedure "D" to estimate the sensible and latent loads produced by ventilation.

5-12 Cooling Equipment Size

Cooling equipment size should be based on the calculated sensible and the calculated latent cooling loads as indicated by calculation Procedure "D". Note that when the cooling equipment is operating at the outdoor design temperature, its sensible capacity must be at least equal to the calculated sensible load and its latent capacity must be equal to or greater than the calculated latent load.

Air conditioning (cooling only) equipment should be selected to keep oversizing to a minimum. (Refer to Section 7-27.)

In order to take full advantage of the heat pump during the heating season, heating and cooling heat pumps can be oversized for cooling by up to 25 percent. (Refer to Section 7-28.)

Multi-zone equipment sizing involves special considerations. (Refer to Appendix 2 for more information on calculating residential zone loads and sizing multi-zone equipment.)

REQUIRED MINIMUM FREE VENT AREA

Gable vents only **without** ceiling vapor barrier

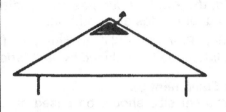

1 sq. ft. for each 150 sq. ft. of ceiling. Put half in each gable end.

Gable vents only **with** vapor barrier in ceiling

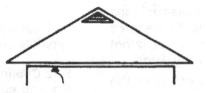

1 sq. ft. for each 300 sq. ft. of ceiling. Put half in each gable end.

Combination of gable/eave vents **without** vapor barrier

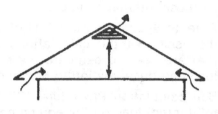

0.5 sq. ft. in each gable end for each 300 sq. ft. of ceiling and 0.5 sq. ft. of vent area in each soffit or at each eave for each 300 sq. ft. of ceiling.

Continuous ridge/eave vent system

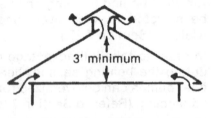

1 sq. ft. at ridge and ½ sq. ft. at each soffit for each 300 sq. ft. of ceiling.

Roof vent/eave vent system

1 sq. ft. at roof and ½ sq. ft. at each soffit for each 300 sq. ft. of ceiling.

Note:

Cathedral, flat and mansard roof/ceilings with vapor barriers need vent areas of 1 sq. ft. per 300 sq. ft. of ceiling. Cross ventilation may be achieved by placing half of the required vent area at each soffit or eave. Experience has shown that if possible, it is preferabe to provide half of the required area through gable, roof, or continuous ridge vents, and one quarter at each soffit or eave. Without a vapor barrier, the vent area should be doubled.

The above stated vent areas for attics and roof/ceilings refer to net free area of opening through which air can pass unobstructed. When screening, louvers, and rain/snow shields cover the vents, the area of the vent opening should be increased to offset the area of the obstructions. For convenience, the table below lists a recognized method of determining gross area of vent opening related to type of vent covering and required net free ventilating area.

Type of Covering	Size of Opening
1/4" hardware cloth	1 x required net free area
1/4" hardware cloth and rain louvers	2 x required net free area
1/8" mesh screen	1-1/4 x required net free area
1/8" mesh screen and rain louvers	2-1/4 x required net free area
1/16" mesh screen	2 x required net free area
1/16" mesh screen and rain louvers	3 x required net free area

Figure 5-1 HUD-FHA Attic Ventilation Standard

SECTION VI
Example Problem: Heat Gain Calculation

The heat gain calculation will be made for the same structure used for the heat loss calculation. In both examples, the measurements and the construction details are the same.

6-1 Design Temperature Differences

The design temperature difference is the air temperature difference across a structural component. Table 1 lists the outside design temperatures for various locations. Assume the inside design temperature to be 75°F, 55% relative humidity. Unconditioned space temperatures should be estimated as close as possible by considering the location and the use of the space in question. The design temperature difference is used to select the appropriate HTM for the structural components.

6-2 Calculation Procedure

Once the area, construction details, and temperature differences are determined, the data in the back of this Manual and the Manual J worksheet can be used to determine the heat gain. The total heat gain is the sum of the heat gains through the building envelope; (solar gain, transmission, infiltration and internal loads). HTM for various structural components are in Table 3 and 4. HTM values for temperature differences which fall between those listed in the table should be interpolated as discussed. Internal gains and infiltration gains should be calculated by using the procedures outlined in Section V.

Data must be obtained from drawings or by a field inspection before load calculations can be made. The data required includes:

A. Measurements to determine areas:
 1. Running feet of exposed wall.
 2. Length and width of rooms and house.
 3. Ceiling heights.
 4. Dimensions of windows and doors.

B. Area calculations:
 1. Gross area of walls exposed to the outdoor conditions.
 2. Gross area of partitions.

 3. Gross area of walls below grade.
 4. Areas of window and doors.
 5. Areas of ceilings under an attic or unconditioned space and/or roof ceiling combinations.
 6. Areas of floors exposed to the outdoors or floors over a unconditioned space or over a crawl space and/or basement floors.
 7. Running feet of exposed perimeter for slab on grade floors.

Closets and halls are usually included with adjoining rooms. Entrance halls should be considered separately. Wall, floor, or ceiling dimensions can be rounded to the nearest foot. Window dimensions are recorded to the nearest inch. Measure the size of window or door opening: do not include frame.

C. Construction Details:
 1. Exposed walls and partitions.
 2. Windows and glass doors.
 3. Panel Doors.
 4. Ceilings and roof-ceilings.
 5. Floors and ground slabs.

D. Temperature differences:
 1. Temperature differences across all components exposed to the outdoor conditions.
 2. Temperature differences across partitions and all floors and ceilings that are adjacent to unconditioned spaces.

6-3 Example

Figures 6-1 and 6-2 represent a house located in Cedar Rapids, Iowa. Figure 6-3 lists the construction details. Assume the inside design temperature is 75°F. From Table 1, the summer design temperature is 88°db, 75°wb, 38 grains moisture difference with a medium (M) daily range. Figure 6-3 shows the completed worksheet. Note that the outside design temperature is rounded from 88°db to 90°db to expedite the calculations. Rounding the design temperature difference up by three degrees or down by one degree will not produce any serious errors in the calculations.

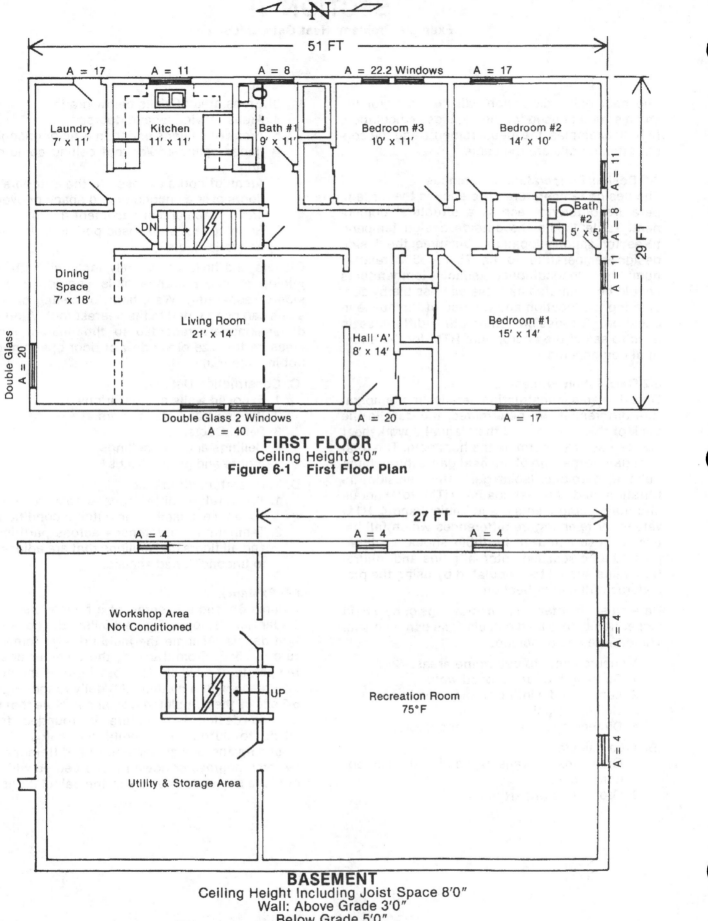

FIRST FLOOR
Ceiling Height 8'0"
Figure 6-1   First Floor Plan

BASEMENT
Ceiling Height Including Joist Space 8'0"
Wall: Above Grade 3'0"
Below Grade 5'0"
Figure 6-2   Basement Plan

FIGURE 6-3 EXAMPLE HEAT GAIN CALCULATION
DO NOT WRITE IN SHADED BLOCKS

					Entire House			1 Living			2 Dining			3 Laundry			4 Kitchen			5 Bath-1		
1	Name of Room				Entire House			Living			Dining			Laundry			Kitchen			Bath-1		
2	Running Ft. Exposed Wall				160			21			25			18			11			9		
3	Room Dimensions Ft.				51 x 29			21 x 14			7 x 18			7 x 11			11 x 11			9 x 11		
4	Ceiling Ht. Ft — Directions Room Faces				8			8 West			8 North			8			8 East			8 East		

	TYPE OF EXPOSURE	Const No.	HTM Htg.	HTM Clg.	Area/Length	Btuh Htg.	Btuh Clg.	Area/Length	Btuh Htg.	Btuh Clg.	Area/Length	Btuh Htg.	Btuh Clg.	Area/Length	Btuh Htg.	Btuh Clg.	Area/Length	Btuh Htg.	Btuh Clg.	Area/Length	Btuh Htg.	Btuh Clg.
5	Gross a	12-d			1280			168			200			144			88			72		
	Exposed b	14-b			480																	
	Walls & c	15-b			800																	
	Partitions d	13N			232																	
6	Windows a																					
	& Glass b																					
	Doors Htg. c																					
	d																					
7	Windows — North			14	20		280				20		280									
	& Glass — E & W			44	115		5060	40		1760							11		484	8		352
	Doors Clg. — South			23	30		690															
	Basement			70/36	8/8		848															
8	Other Doors	11-e		3.5	37		130							17		60						
9	Net a	12-d		1.5	1078		1617	128		192	180		270	127		191	77		116	64		96
	Exposed b	14-b		1.6	233		373															
	Walls & c	15-b		0																		
	Partitions d	13-n		0																		
10	Ceilings a	16-d		2.1	1479		3106	294		617	126		265	77		162	121		254	99		208
	b																					
11	Floors a	21-a		0																		
	b	19-f		0																		
12	Infiltration HTM			7.18	218		1565	40		287	20		144	17		122	11		79	8		57
13	Sub Total Btuh Loss = 6+8+9+10+11+12																					
14	Duct Btuh Loss		%																			
15	Total Btuh Loss = 13+14																					
16	People @ 300 & Appliances 1200						3000	3		900	3		900			—			1200			—
17	Sensible Btuh Gain = 7+8+9+10+11+12+16						16669			3756			1859			535			2133			713
18	Duct Btuh Gain		%				—			—			—			—			—			—
19	Total Sensible Gain = 17+18						16669			3756			1859			535			2133			713

NOTE: USE CALCULATION PROCEDURE D TO CALCULATE THE EQUIPMENT COOLING LOADS

*Answer for "Entire House" may not equal the sum of the room loads if hall or closet areas are ignored or if heat flows from one room to another room.

From Table 4

		Const. No.	HTM
	ASSUMED DESIGN CONDITIONS AND CONSTRUCTION (Cooling)		
A.	Outside Design Temperature: Dry Bulb 88 Rounded to 90 db 38 grains - Table 1		
B.	Daily Temperature Range: Medium - Table 1		
C.	Inside Design Conditions: 75F, 55% RH Design Temperature Difference =	(90-75 = 15)	
D.	Types of Shading: Venetian Blinds on All First Floor Windows - No Shading, Basement		
E.	Windows: All Clear Double Glass on First Floor - Table 3A		
	North		14
	East or West		44
	South		23
	All Clear Single Glass (plus storm) in Basement - Table 3A Use Double Glass		
	East		70
	South		36
F.	Doors: Metal, Urethane Core, No Storm, 0.50 CFM/ft.	11e	3.5
G.	First Floor Walls: Basic Frame Construction with ½" Asphalt Board (R-11) - Table 4	12d	1.5
	Basement Wall: 8" Concrete Block, Above Grade: 3 ft (R-5) - Table 4	14b	1.6
	8" Concrete Block Below Grade: 5 ft (R-5) - Table 4	15b	0
H.	Partition: 8" Concrete Block Furred, with Insulation (R-5), Δ T approx. 0°F - Table 4	13n	0
I.	Ceiling: Basic Construction Under Vented Attic with Insulation (R-19), Dark Roof - Table 4	16d	2.1
J.	Occupants: 6 (Figured 2 per Bedroom, But Distributed 3 in Living, 3 in Dining)		
K.	Appliances: Add 1200 Btuh to Kitchen		
L.	Ducts: Located in Conditioned Space - Table 7B		
M.	Wood & Carpet Floor Over Unconditioned Basement, Δ T approx. 0°F	19	0
N.	The Envelope was Evaluated as Having Average tightness - (Refer to the Construction details at the Bottom of Figure 3-3)		
O.	Equipment to be Selected From Manufacturers Performance Data.		

	6 Bedroom 3			7 Bedroom 2			8 Bath 2			9 Bedroom 1			10 Hall			11 Rec. Room			12 Shop & Utility			1
	10			24			5			29			8			83			88			2
	10 x 11			14 x 10			5 x 5			15 x 14			8 x 14			27 x 29			24 x 29			3
	8 East			8 E & S			8 South			8 S & W			8 West			8 E & S			8 East			4
	Area or Length	Htg	Clg	Area or Length	Htg	Clg	Area or Length	Htg	Clg	Area or Length	Htg	Clg	Area or Length	Htg	Clg	Area or Length	Htg	Clg	Area or Length	Htg	Clg	
	80			192			40			232			64			249			231			5
																415			385			
																232						
																						6
	22		968	17		748				17		748				8/8		560				7
				11		253	8		184	11		253				8/8		288				
													20		70							8
	58		87	164		246	32		48	204		306	44		66							9
																233		373				
	110		231	140		294	25		53	210		441	112		236							10
																						11
	22		158	28		201	8		57	28		201	20		144	16		115				12
																						13
																						14
																						15
			—			—			—			—			—			—				16
			1444			1742			342			1949			516			1336				17
			—			—			—			—			—			—				18
			1444			1742			342			1949			516			1336				19

Line 1. Identify each area.

Lines 2 and 3. Enter the pertinent dimensions from Figures 6-1 and 6-2.

Line 4. For reference, enter the ceiling height and the direction the glass faces.

Lines 5A through 5D. Enter the gross wall area for the various walls. For rooms with more than one exposure, use one line for each exposure. For rooms with more than one type of wall construction, use one line for each type of construction. Find the construction number in the tables in back of this manual. Enter the construction number on the appropriate line.

Example: The gross area of the west living room wall is 168 sq. ft. This wall is listed in Table 4, number 12, line D. The construction number is 12-d.

Line 6. Not required for cooling calculations.

Line 7. Enter the areas of windows and glass doors for the various rooms and exposures. Use the drawings and construction details, or determine by inspection, the types of windows used in each room. Also note the shading and the exposure. Refer to the tables in the back of this manual and select the HTM for each combination of window, shading and exposure. Enter the HTM values in the column designated cooling. Multiply each window area by its corresponding HTM to determine the heat gain through the window. Enter this value in the column Btuh - Cooling.

Example: The living room has 40 sq. ft. of west-facing glass. The window is double pane with drapes or blinds. The design temperature difference is rounded to 15°F. The HTM listed in Table 3A, (double glass, drapes, or venetian blinds, design temperature difference of 13°F), is Btu/(hr. sq. ft.) The heat gain is:

44 Btu/(hr. sq. ft.) x 40 sq. ft. = 1,760 Btuh.

Example: The dining area has 20 sq. ft. of north-facing glass. The HTM listed in Table 3A, (double glass, draperies, 15°F temperature difference) is 14 Btu/(hr. sq. ft.) The heat gain is:

14 Btu/(hr. sq. ft.) x 20 sq. ft. = 280 Btuh.

Example: The glass in the basement recreation room is single pane with storm. The design temperature difference is 15°F. The room has 8 sq. ft. of east-facing glass and 8 sq. ft. of southfacing glass. From Table 3A, (double pane, clear glass, 15°F design temperature difference), the HTM are 70 Btu/(hr. sq. ft.) for the east and 36 Btu/(hr. sq. ft.) for the south. The heat gian for the east window is:

70 Btu/(hr. sq. ft.) x 8 sq. ft. = 560 Btuh.

The heat gain for the south window is:

36 Btu/(hr. sq. ft.) x 8 sq. ft. = 288 Btuh.

Line 8. For each room, enter the area of any doors that are not glass. From Table 4, No. 11-E, select the HTM and enter this value on the worksheet. The heat gain through the door is calculated by multiplying the HTM by the area of the door. Enter the heat gain in the appropriate column. Laundry door 17 sq. ft. x 3.5 Btu/(hr. sq. ft.) = 60 Btuh.

Line 9A through 9D. For each room, subtract the window and door areas from the corresponding gross wall area and enter the net wall areas and corresponding construction numbers. From Table 4, select the HTM for each wall. Multiply the HTM by the appropriate net wall area and enter the heat gain through the wall.

Example: The west wall in the living room has a net area of (168 sq. ft. - 40 sq. ft.) = 128 sq. ft. The design temperature difference is rounded to 15°F and the daily range is M. From Table 4, No. 12-D, the HTM is 1.5 Btu/(hr. sq. ft.) The heat gain through the living room wall is:

1.5 Btu/(hr. sq. ft.) x 128 sq. ft. = 192 Btuh.

Example: The basement wall in the recreation room has a net area of 233 sq. ft. above grade and 410 sq. ft. below grade. From Table 4, No. 14-B, the HTM is 1.6 Btu/(hr. sq. ft.) The heat gain through the above grade wall is:

1.6 Btu/(hr. sq. ft.) x 233 sq. ft. = 373 Btuh.

(Wall below grade need not be included in the heat gain calculation).

Lines 10A and 10B. Enter the ceiling area for the various rooms and the construction number for the corresponding ceiling. Determine the HTM from Table 4 and enter it. Multiply the HTM by the ceiling area and enter the heat gain.

Example: The living room ceiling has an area of 294 sq. ft. The (dark roof) construction number is (16-D), the design temperature difference is 15, and the daily range is M. From Table 4, No. 16-D, the HTM is 2.1 Btu/(hr. sq. ft.). The heat gain through the living room ceiling is:

2.1 Btu/(hr. sq. ft.) x 294 sq. ft. = 617 Btuh.

Lines 11A and 11B. For a room which will experience a gain through the floor, enter the floor area and the corresponding construction number. Determine the HTM from Table 4 and enter it. Multiply the HTM by the appropriate area and enter the heat gain.

Line 12. Use Table 5 and calculation Procedure B to calculate the summer infiltration HTM, and enter this value on Line 12. For each room, enter the total sq. ft. of the window and door openings on Line 12. Finally, compute the infiltration heat gain due to infiltration for each room by multiplying the summer infiltration HTM value by the appropriate sq. ft. of window and door opening. Enter these results on Line 12. Refer to Figure 6-4 for an example of this calculation.

Lines 13 through 15. These lines are not used for the cooling calculation.

Line 16. Enter sensible internal loads due to appliances and occupants for the rooms.

Line 17. For each room add all the cooling loads (7, 8, 9, 10, 11, 12 and 16) and enter the totals on line 17.

Line 18. If the duct system is installed in an unconditioned space, enter an allowance for the duct gain for each room on line 18. Refer to Table 7-B for the duct gain multipliers.

Line 19. For the entire house and for each room, add line 17 (structure gain), to line 18 (duct gain) and enter the space sensible gain on the form.

Note: Line 19 on the form only provides information on the space sensible loads. Equipment selection requires calculation of the space latent loads, and if ventilation is used, the sensible and latent ventilation loads must also be calculated.

Calculation Procedure C. Use Calculation Procedure C to calculate the latent infiltration load for the entire house. Refer to Figure 6-5 for an example of the Procedure C calculation.

Calculation Procedure D. Use Calculation Procedure D to estimate the sensible and latent ventilation loads and to estimate the latent loads that are produced by internal sources. Also use Calculation Procedure D to estimate the total sensible and the total latent loads that must be satisfied by the cooling equipment. Refer to Figure 6-5 for an example of these calculations. Equipment to be selected from manufacturers performance data, use RSM = 1.0 (Refer to Table 6).

Table 5

Infiltration Evaluation

Summer air changes per hour

Floor Area	900 or less	900 - 1500	1500 - 2100	over 2100
Best	0.2	0.2	0.2	0.2
Average	0.5	0.5	0.4	0.4 *
Poor	0.8	0.7	0.6	0.5

Envelope Evaluation
Average - Plastic vapor barrier, major cracks and penetrations sealed, tested leakage of windows and doors between 0.25 and 0.50 CFM per running foot of crack, electrical fixtures which penetrate the envelope not taped or gasketed, vents and exhaust fans dampered, combustion air from indoors, intermittent ignition and flue damper, some duct leakage to unconditioned space.

Procedure B - Summer Infiltration HTM Calculation

1. Summer Infiltration CFM
   0.40 AC/HR x 14181 Cu. FT. x 0.0167   =   94.7 CFM
   Volume

2. Summer Infiltration Btuh
   1.1 x 94.7 CFM x 15 Summer TD   =   1563 Btuh

3. Summer Infiltration HTM
   1563 Btuh ÷ 218 Total Window   =   7.17 HTM
        & Door Area

* Includes Finished Basement

Above grade volume = (51x29)x8 + (27x29)x3 = 14181

Figure 6-4 Example Summer Infiltration Calculation

Calculation Procedure C · Latent Infiltration Gain For The Entire House

0.68 x 38 gr diff x 95 CFM	= 2455 Btuh

Calculation Procedure D · Equipment Sizing Calculation · No Mechanical Ventilation

1. Sensible Sizing Load

   Sensible Ventilation Load
1.1 x 0 Vent CFM x / Summer TD	=	0	Btuh
Sensible Load For Structure (Line 19)	+	16669	Btuh
Sum of Ventilation and Structure Loads	=	16669	Btuh
Rating & Temperature Swing Multiplier	=	x 1.0	RSM
Equipment Sizing Load · Sensible	=	16669	Btuh

2. Latent Sizing Load

   Latent Ventilation Load
0.68 x 0 Vent CFM x / gr diff	=	0	Btuh
Internal Loads = 230 x 6 No. People	+	1,380	Btuh
Infiltration Load From Procedure C	+	2,455	Btuh
Equipment Sizing Load · Latent	=	3,835	Btuh

If 150/CFM of ventilation is included with this problem the Table D calculation would appear as shown below.

Calculation Procedure D · Equipment Sizing Calculation

1. Sensible Sizing Load

   Sensible Ventilation Load
1.1 x 150 Vent CFM x 15 Summer TD	=	2475	Btuh
Sensible Load for Structure (Line 19)	+	16669	Btuh
Sum of Ventilation and Structure Loads	=	19144	Btuh
Rating & Temperature Swing Multiplier	x	1.0	RSM
Equipment Sizing Load · Sensible	=	19144	Btuh

2. Latent Sizing Load

   Latent Ventilation Load
0.68 x 150 Vent CFM x 38 gr diff	=	3876	Btuh
Internal Loads = 230 x 6 No. People	+	1380	Btuh
Infiltration Load From Procedure C	+	2455	Btuh
Equipment Sizing Load · Latent	=	7711	Btuh

Figure 6-5 Example Heat Gain Calculation

FORM J—1
Including Calculation Procedures A, B, C, D
Copyright by the
Air Conditioning
Contractors of America
1513 16th Street N.W.
Washington, D.C. 20036
Printed in U.S.A.
1986

| Plan No. _____ |
| Date _____ |
| Calculated by _____ |

WORKSHEET FOR MANUAL J

LOAD CALCULATIONS FOR RESIDENTIAL AIR CONDITIONING

For: Name _Example Problem_____

Address _____

City and State or Province _____

By: Contractor_____

Address _____

City _____

Design Conditions

Winter

Outside db ___-5___ °F Inside db ___70___ °F

Winter Design Temperature Difference ___75___ °F

Summer

Outside db ___88___ °F Inside db ___75___ °F

Summer Design Temperature Difference ___15___ °F

Room RH ___55%___ Daily Range ___M___

Heating Summary

Total Heat Loss for Entire House (Line 15) = ___47,808___ Btuh

Ventilation CFM = ___none___ Winter Design Temperature Difference = _____ °F

Heat Required for Ventilation Air = 1.1 X _____ CFM X _____ °F = ___0___ Btuh

Design Heating Load Requirement = ___47,808___ (house) ___0___ (Vent) = ___47,808___ Btuh

Cooling Summary

Total Sensible Gain ___16,669___ Btuh (Calculation Procedure D) Design Temperature Swings

Total Latent Gain + ___3,835___ Btuh (Calculation Procedure D) Normal 3° (X) 4.5° ( )

Total = Sens. + Lat. = ___20,504___ Btuh Ventilation CFM = ___none___

Equipment Summary

Make _____ Model _____ Type _____

Heating Input (Btuh) _____ Heating Output (Btuh) _____ Efficiency _____

Sensible Cooling (Btuh _____ Latent Cooling (Btuh) _____ Total (Btuh) _____

COP/EER/SEER/HSPF _____ Cooling CFM _____ Heating CFM _____

Space Thermostat Heat ( ) Cool ( ) Heat/Cool ( ) Night Setback ( )

Construction Data

Windows _____

_____

_____

Doors_____

_____

Walls_____

_____

_____

Roof _____

Ceiling _____

Floor _____

_____

Partitions _____

_____

Basement Walls _____

_____

_____

Ground Slab_____

_____

Figure 6-6 Example Load Calculation Summary

SECTION VII
Basic Principles

7-1 Heat

All objects contain heat. If an object or substance were devoid of heat, it would have to be approximately 460°F below zero. The amount of heat contained in an object or substance depends on:

A. The temperature of the object.
B. The weight of the object (obviously 10 lbs. of steel at 200°F contains more heat than 5 lbs. of steel at 200°F).
C. The specific heat of the substance. Specific heat is a property of the substance and can be measured in the laboratory.

For comparison, the specific heat of some familiar substances are:
Standard air: .... 0.24 Btu/lb. °F
Stone, Concrete, Brick: .... 0.20 Btu/lb. °F
Wood: ...... varies between 0.45 - 0.65 Btu/lb. °F
Water: .... 1.00 Btu/lb. °F

7-2 Btu

The Btu (British Thermal Unit) is the unit used to express the amount of heat held in an object or the amount of heat transferred from one substance to another. Water is used as a reference. A Btu is defined as the amount of heat required to raise the temperature of one pound of water one degree Fahrenheit.

7-3 Heat Transfer

Objects or substances at higher temperatures lose or give up some of their heat to objects or substances at lower temperatures. The amount of heat an object gains or loses can be calculated if the change in temperature of the object is known.

Let Q represent the heat gain or loss in Btu, then:
$Q = W \times SpH \times TD$.

Where:
W is the weight of the substance in pounds.
SpH is the specific heat of the substance -Btu/lb. °F.
TD is the temperature change in °F.

Example: The amount of heat required to raise the temperature of 10 lbs. of air, (C = 0.24 Btu/lb. °F) from 20° to 90°F is:

Q (Btu) = 10 (lbs.) x 0.24 (Btu/lb. °F) x (90-20) (°) = 168 Btu.

In heating and air conditioning work, the formula for heating and cooling air is written as:

$Q = 1.1 \times CFM \times TD$.

Where:
Q is the sensible heat removed from or added to the air (Btuh).
CFM is the (cu. ft./min.) of air that is heated or cooled.
TD is the change in temperature of the air (°F).
1.1 is a constant (that includes the effect of the specific heat of air) which relates the air flow (in cubic feet per minute) and temperature change to the heat transfer, (in Btuh). (60 min./hr. x 0.076 lb./cu. ft. x 0.24 Btu/lb. °F = 1.1 Btuh/cfm. °F).

Example: If 135 CFM of air is heated from 20°F to 90°F, the required heat input is:

Q (Btuh) = 1.1 x 135 CFM x (90-20) °F = 10,400 Btuh.

7-4 Modes of Heat Transfer

In air conditioning and heating, the goal is to **prevent** the inside of the structure from becoming too warm in summer or too cold in winter. If the inside temperature is to remain constant, the cooling and heating systems must remove the heat that comes into the structure or replace the heat that is lost from the structure. Since heat is continuously transferred, we must know how much heat is transferred in a given time period. One hour is commonly used, so heat gains or losses are calculated in Btuh.

Conduction:

Heat is transferred **through** a substance by conduction. The heat flow depends on the temperature difference between the hot and cold surfaces, the area, the thermal conductivity, and the thickness of the material. The thermal conductivity of a substance is a property which can be measured in the laboratory. Conductivities are usually specified for one-inch of thickness. If the temperature difference across the object is held constant, there will be a steady flow of heat through the object.

Let Q represent the heat flow in Btuh, then:

$$Q = \frac{K \times A \times TD}{L}$$

Where:
K is the conductvity in (Btu in.)/(hr. sq. ft. °F).
A is the area in square feet.
TD is the temperature difference in °F.
L is the thickness of the material in inches.

Example: A solid 4 inch brick wall is 120°F on one side and 70°F on the opposite side. The conductivity for **one inch** is 5.0 (Btu in.)/(hr. sq. ft. °F). The heat flow through one sq. ft. of wall is:

$$Q \text{ (Btuh)} = \frac{5.0 \text{ (Btuh in./Sq. ft.)} \times 1 \text{ (sq. ft.)} \times (120-70) \text{ (°F)}}{4 \text{ (in.)}}$$

$$= 62.5 \text{ (Btuh)}$$

Convection:

When heat is carried to an object or away from an object by a fluid, the heat is transferred by convection. The heat exchanger in a home furnace is an example. The air is forced over the hot heat exchanger surfaces and carries the heat to the occupied space.

Calculation of heat transferred by convection requires considerable engineering background. The basic equation which determines heat flow due to convection is:

$$Q = H \times A \times TD$$

Where:

Q is the heat flow in Btuh.

A is the area of the heat transfer surface in sq. ft.

TD is the temperature difference between the air and the heat transfer surface in °F.

H is the convection coefficient in Btu/(hr. sq. ft. °F).

The problem is that "H" is very difficult to determine. There are methods, however, for the layman to use in estimating convective heat transfer. In making load calculations, convection will affect heat transfer through walls, roofs and glass.

Radiation:

Radiation is heat carried to or away from an object by electromagnetic waves. It differs from convection because the heat can be transferred from one object to another without the help of a fluid. The interaction of the sun and the earth is an example of heat transfer through radiation. Heat transfer by radiation for the most part is unaffected by the earth's atmosphere. Heat is transferred through the air, but the air is not heated by the electromagnetic waves. Radiation must strike a solid object to convert electromagnetic waves into heat. The solid object is heated and then releases this heat by convection. Radiant heaters used in garages are an example. The basic equation which determines the heat flow due to radiation is not included here, but methods which allow the effects of the radiation from the sun to be accounted for when the cooling load is computed, are available.

7-5 Conductivity

Conductivity is a material property. Materials that are good conductors of heat, such as metals, have high conductivities. Insulators, such as fiber glass or asbestos, have low conductivities. Conductivities are measured in the laboratory and are usually listed for one-inch thick material.

7-6 Conductance

Conductance is closely related to conductivity. The conductivity of a material is specified for one inch of material. The conductance of a material is specified when the thickness is other than one-inch.

Example: 1-inch of glass fiber insulating board has a conductivity of 0.25 Btu/(hr. sq. ft. °F) per inch. A 3-inch thick board would have a conductance of 0.083 Btu/(hr. sq. ft. °F).

7-7 Resistance Per Inch and Resistance

Resistance per inch is the reciprocal of the conductivity; and the resistance is the reciprocal of the conductance.

Example: 1 inch of insulating board has a conductivity of 0.25 Btu/(hr. sq. ft. °F) and a resistance per inch of 4.0 (sq. ft. °F hr)/Btu. 3 inches of insulating board has a conductance of 0.083 Btu/(hr. sq. ft. °F) and a resistance of 12.0 (sq. ft. °F hr.)/Btu.

7-8 Resistance of a Composite Structure

The resistance of a composite wall can be calculated by adding the resistances of the individual components.

Example: A 2 x 4 stud wall with brick veneer, 3 inches of insulation and a dry wall interior would have a combined resistance of 12.89 (sq. ft. °F hr.(/Btu.

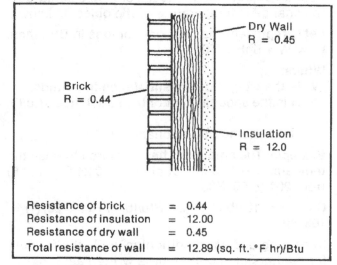

Resistance of brick	= 0.44
Resistance of insulation	= 12.00
Resistance of dry wall	= 0.45
Total resistance of wall	= 12.89 (sq. ft. °F hr)/Btu

7-9 R Values

The R value is the thermal resistance of a given thickness of insulating material. For instance, 6 inches of fiberglass blanket has a resistance of 19 (sq. ft. °F hr.)/Btu and 3 inches of Urethane also has a resistance of 19 (sq. ft. °F hr.)/Btu. Both the Urethane and the glass blanket are rated R-19. Figure 7-1 lists R values for various types of insulation. When insulation is compressed to less than its normal thickness, it loses some of its ability to insulate and its R value decreases because the amount of air trapped in the material is reduced. Figure 7-2 summarizes the effect of compression on R values. Table 10 lists R values for common building materials.

7-10 Air Film Coefficients

Heat transferred through a wall must include the effect of convection at the inside and outside surfaces. Air film coefficients account for convection; (refer to 7-4, "convection"). The air film resistance which should be added to the wall resistance to account for convection is given in Figure 7-3 "Wall Air Film Resistance" on page 46.

	Approx. R/Inch	Approximate Inches Needed For					
		R11	R19	R22	R30	R38	R49
Loose Fill Machine-blown Fiberglass	R2.25	5	8.5	10	13.5	17	22
Mineral Wool	R3.125	3.5	6	7	10.0	12.5	16
Cellulose	R3.7	3	5.5	6	8.5	10.5	13.5
Loose Fill Hand-poured Cellulose	R3.7	3	5.5	6	8.5	10.5	13.5
Mineral Wool	R3.125	3.5	6	7	10.0	12.5	16
Fiberglass	R2.25	5	8.5	10	13.5	17	22
Vermiculite	R2.1	5.5	9	10.5	14.5	18	23.5
Batts or Blankets Fiberglass	R3.14	3.5	6	7	10.0	12.5	16
Mineral Wool	R3.14	3.5	6	7	10.0	12.5	16
Rigid Board Polystyrene beadboard	R3.6	3	5.5	6.5	8.5	10.5	14
Extruded Polystyrene	R4-5.41	3-2	5-3.5	5.5-4	7.5-5.5	9.5-7	12.5
(Styrofoam) Urethane	R5.4-6.2	2	3	3.5	5.0	6.5	8
Fiberglass	R4.0	3	5	5.5	7.5	9.5	12.5

Figure 7-1 R Values For Insulation

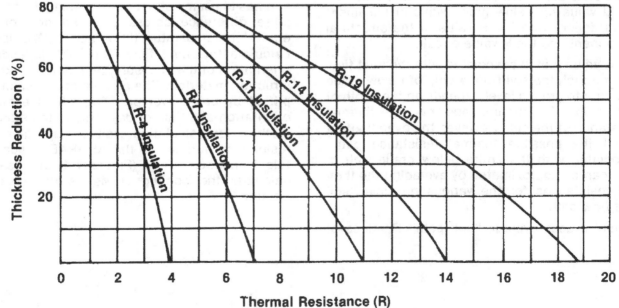

Figure 7-2 Effect of Compression on R Values

Figure 7-3
Wall Air Film Resistance (sq. ft. °F hr.)/Btu

	Summer	Winter
	7.5 mph wind	15 mph wind
Outside Surface	0.25	0.17
Inside Surface	0.68	0.68

Example: The wall discussed in 7-8 had a total resistance for brick insulation, and plaster board of 12.89 (sq. ft. °F hr.)/Btu. The inside and outside air films will increase the wall resistance. For the summer:

Outside air film resistance = 0.25
Wall resistance = 12.89
Inside air film resistance = 0.68
Net effective resistance = 13.82 (sq. ft. °F hr.)/Btu.

7-11 Transmission Coefficients (U)
The U value combines the effect of the thermal resistance of the wall, ceiling/roof, or glass, and the effects of convection at the inside and outside surfaces. U values are expressed in Btuh/sq. ft. °F and are tabulated for all common types of construction. The U value for any type of construction can be calculated by computing the reciprocal of the overall resistance (including air film resistance) of the wall, roof, or glass.

Example: in 7-10, the net effective resistance of the wall was 13.82 (sq. ft. °F hr.)/Btu. the transmission coefficient is found by taking the reciprocal of this number.

U = 1.0/13.82 = 0.072 Btu/(hr. sq. ft. °F).

7-12 Below Grade Transmission Coefficients
The U value for below grade walls or basement floors depends on the R value of the structural component and the R value of soil.

The R value that is associated with the heat flow path through the soil from a strip of below grade structure to ground level depends on the depth of that strip of structure and on the thermal resistance of the soil. However, the overall effect of all the possible thermal resistance paths associated with the entire below grade wall or floor can be approximated by averaging the thermal resistances for the various possible "soil heat flow paths."

Manual J uses average "soil path" R values of 4.85 for walls which extend from two to five feet below grade (construction numbers 15 a,b,c,d) and 7.84 for walls which extend from five to eight feet below grade (construction numbers 15 e,f,g,h). No "soil path" R value is applied to walls which are less than two feet below grade (construction number 14). The transmission coefficient for basement floors (construction number 21) also includes an allowance for the "soil path" R value.

7-13 Temperature Difference (TD)
The temperature difference (TD) across the building envelope (walls, ceiling/roof, glass) is the difference between the outside temperature and the room temperature.

Example: Atlanta has a winter design temperature of 22°F (see Table 1). If the room temperature is 70°F, the temperature difference across the building envelope is (70° - 22°) = 48°F.

7-14 Basement Temperature Difference
The average temperature near the surface of the ground during the heating season is somewhat higher than the winter outdoor design temperature. Therefore, the effective temperature difference associated with a below grade wall or basement floor is less than the winter design temperature difference. Manual J applies a adjustment factor of 0.85 to the below grade wall and basement floor U values to account for the benefit of the heat storage capability of the soil.

7-15 Equivalent Temperature Difference (ETD)
The effects of the sun and thermal storage must be included when the heat gain through a wall, roof, or other component is calculated. Since the sun has the same effect as increasing the temperature difference across the wall or roof, the equivalent temperature difference (ETD) is used to calculate this heat flow. The ETD depends on orientation of the exposure, time of day, and construction materials. The procedure has been simplified by using average ETDs which eliminate orientation and time of day. This is the standard method for calculating wall or roof gains for residences outlined in the ASHRAE Fundamentals Handbook. ETDs (Figure 7-4) are used to calculate the cooling HTMs as discussed in Section 7-17.

Design Temperature Difference, deg. F	10		15			20			25		30	35
Daily Temperature Range	L	M	L	M	H	L	M	H	M	H	H	H
WALLS AND DOORS												
1. Frame and veneer-on-frame	17.6	13.6	22.6	18.6	13.6	27.6	23.6	18.6	28.6	23.6	28.6	33.6
2. Masonry walls, 8-in. block or brick	10.3	6.3	15.3	11.3	6.3	20.3	16.3	11.3	21.3	16.3	21.3	26.3
3. Partitions, frame	9.0	5.0	14.0	10.0	5.0	19.0	15.0	10.0	20.0	15.0	20.0	25.0
masonry	2.5	0	7.5	3.5	0	12.5	8.5	3.5	13.5	8.5	13.5	18.5
4. Wood doors	17.6	13.6	22.6	18.6	13.6	27.6	23.6	18.6	28.6	23.6	28.6	33.6
CEILINGS AND ROOFS												
1. Ceilings under naturally vented attic or vented flat roof — dark	38.0	34.0	43.0	39.0	34.0	48.0	44.0	39.0	49.0	44.0	49.0	54.0
— light	30.0	26.0	35.0	31.0	26.0	40.0	36.0	31.0	41.0	36.0	41.0	46.0
2. Built-up roof, no ceiling — dark	38.0	34.0	43.0	39.0	34.0	48.0	44.0	39.0	49.0	44.0	49.0	54.0
— light	30.0	26.0	35.0	31.0	26.0	40.0	36.0	31.0	41.0	36.0	41.0	46.0
3. Ceilings under unconditioned rooms	9.0	5.0	14.0	10.0	5.0	19.0	15.0	10.0	20.0	15.0	20.0	25.0
FLOORS												
1. Over unconditioned rooms	9.0	5.0	14.0	10.0	5.0	19.0	15.0	10.0	20.0	15.0	20.0	25.0
2. Over basement, enclosed crawl space or concrete slab on ground	0	0	0	0	0	0	0	0	0	0	0	0
3. Over open crawl space	9.0	5.0	14.0	10.0	5.0	19.0	15.0	10.0	20.0	15.0	20.0	25.0

Figure 7-4 Equivalent Temperature Differences (ETD).

7-16 Daily Range

The daily range is the average difference between the daily high and the daily low temperatures for a given location. During the night, structures which are located in areas that have a high daily range (large difference in daily outdoor temperature) will be cooled down more than identical structures which are located in areas that have a low daily range. The amount of night time cooling has a direct effect on the heat gain during the following day. The ETD values listed in Figure 7-4 include an allowance for the effect of daily range.

Example: The summer design temperature difference in a city is 90°F-(M). If the room temperature is 75°F, the temperature difference is (90°-75°) = 15°F. Figure 7-4 lists the equivalent temperature differences listed by ASHRAE Fundamentals Handbook for various types of construction.

A frame wall has an ETD of 18.6°F which indicates that the sun striking the wall produces a 3.6°F increase in temperature difference across the wall.

Note that a masonry wall has an ETD of only 11.3°F, which is 3.7°F lower than the actual temperature difference across the wall. The reason is that masonry walls can store a considerable amount of heat which reduces the rate of heat flow into the room. (The effect of the sun is still included in the 11.3°F figure).

A dark colored roof has an ETD of 39.0°F. The roof ETD is very large because the roof is exposed to the sun all day.

7-17 Heat Transfer Multiplier

The heat transfer multiplier (HTM) is the amount of heat that flows through one square foot of building envelope at a given temperature difference.

Heating HTMs for glass, doors, walls, roofs, ceilings and floors (Table 2) are determined by multiplying the transmission coefficient (U) by the winter design temperature difference.

HTM = U value x temperature difference

Cooling HTM multipliers for doors, walls, roofs, ceilings and floors (Table 4) are determined by multiplying the transmission coefficient (U) by the summer equivalent temperature difference. Note that the cooling HTM values must also be determined for the daily range.

HTM = U value x ETD

Cooling HTM multipliers for glass (Table 3) are determined as described in the following paragraph.

7-18 Heat Gains through Glass

In summer, the total heat gain through glass is equal to the combined effects solar radiation and transmission. The solar radiation gain varies with the month, time of day, direction the window is facing (exposure), number of panes and type of glass, type of outdoor and indoor shading, and the capacity of the building materials to store heat. The transmission gain depends on the air temperature difference across the glass and the window or (glass) door U value.

Table 3 in back of this manual can be used to

estimate the total heat gain (solar plus transmission) through glass. The HTM values in this table are based on the average heat gain through glass that occurs in the warmest summer month over an hourly period that extends from mid morning through late afternoon. (These average values are the standard values recommended by the ASHRAE Fundamentals Handbook.) The use of average values eliminates the month, time of day and the heat storage capacity of the building materials from consideration. However Table 3 does require the user to consider exposure, number of panes, type of glass, indoor shading and outdoor shading.

Table 3A lists the cooling HTM values for single, double or triple; (clear, tinted and heat reflective) glass, with or without a low emittance coating, that have various types of internal shading. This table also lists the HTM values for glass that is shaded by awnings, porches, and over-hangs. (Note that the HTM values for external shading are equal to the HTM values which are listed for north facing glass.)

Tables 3B through 3E list the cooling HTM values for glass which is equipped with external shade screens. These HTM values are listed for shade screens which have shading coefficients of 0.15, 0.25, 0.35 and 0.45. The values listed in the tables were calculated by dividing the Table 3A HTM values into their solar and transmission components and multiplying the solar component by a shading coefficient and then adding the heat gain due to transmission to this product. All the same combinations of construction, internal shading and glass type that were listed in Table 3A are listed in these tables.

Table 3F lists the cooling HTM values for single or double pane; (clear, tinted and reflective) sky-lights. The values are listed for various exposures and for various angles of inclination. The tables were developed by summing the heat gains through the projected areas (vertical and horizontal) of one square foot of sky-light.

Use Table 8 to determine the area (square feet) of shaded and unshaded glass which is associated with horizontal projections. The heat gain for the shaded area will be equal to the heat gain for the same area of north glass. The heat gain for the unshaded area will be equal to the heat gain for the same area of glass which faces in the direction of the exposure in question. The total heat gain will be the sum of the gains for the shaded and unshaded areas.

7-19 External Shading Coefficient
The external shading coefficient is defined as the ratio of the solar gain for a single pane of clear

glass which is equipped with an external shade screen but which does not have internal shading to the solar gain for a reference glass. The reference glass consists of a single pane of clear glass that is not shaded at all. Shading coefficients are determined by laboratory test.

7-20 Infiltration
Infiltration is defined as the uncontrolled leakage of outdoor air into the occupied space through openings in the building envelope. This leakage adds to the room heating and cooling loads of the structure. Note that some infiltration is necessary to maintain an acceptable level of air quality or to provide (in some cases) combustion air for a furnace. Refer to Table 9 for a procedure which can be used to determine if infiltration can meet the fresh air requirements of the structure.

7-21 Ventilation
Ventilation is defined as outdoor air that is introduced through the HVAC equipment in a controlled manner. Ventilation will increase the heating and cooling load on the equipment but will have no effect on the heating or cooling loads within the structure. Normally, mechanical ventilation is not used in residential heating and cooling systems. However, mechanical ventilation may be necessary for those structures that are so air tight that the infiltration rate is insufficient for dilution of odors, gases and smoke, winter humidity control, make-up air for exhaust fans or for combustion air. Refer to Table 9 for a procedure which can be used to determine if ventilation is required.

The amount of energy that is required to condition ventilation air can be reduced if an air to air heat exchanger is used to reclaim heat from air which is exhausted from the structure in a controlled manner.

7-22 Sensible Heat
The heat associated with temperature change that occurs when a structure loses or gains heat is called sensible heat. Changes in sensible heat are easily detected by a dry bulb thermometer.

7-23 Latent Heat
The heat associated with the amount of moisture that the structure gains or loses is called latent heat. Changes in latent heat appear as changes in relative humidity and are detected by a humidistat. Heat is required to evaporate water into the air. Conversely, when moisutre in the air is condensed to water, heat is liberated. Since the heat added or removed from the air results in a change of state (water to vapor or vapor to water) the dry bulb temperature does not change when latent heat is gained or lost.

7-24 Internal Loads

Sources of heat contained within the structure; lights, people, motors, and cooking equipment are examples of internal loads.

7-25 Room Load

The heat loss or gain calculated for one room of a structure is called the room load.

7-26 Rating and Swing Multiplier

When operating at the design conditions, residential cooling equipment is normally sized to limit the deviation from the indoor design temperature to 3° fahrenheit or less. A correction factor must be applied to the design cooling load if the cooling equipment is selected for a temperature swing that exceeds 3°F. Table 6 lists factors which can be used to size equipment for a 4.5 degree swing. (ACCA does not recommend sizing for temperature swings that exceed 4.5 degrees).

No load correction is required for a three degree swing when cooling equipment is selected from the manufacturers performance data at the same indoor and outdoor design temperatures that were used for the load calculations. However, a 0.90 correction is required if a 4.5 degree swing is desired. Refer to Table 6.

When cooling equipment is selected from the "standard ratings" (which are listed in the ARI Directory), the design load must be adjusted to match the outdoor temperature (95°F) that was used to establish the published ARI capacities. Table 6 lists the correction factors (for either a 3 or 4.5 degree swing) which must be applied to the calculated design load when the outdoor design temperature is not equal to 95°F.

7-27 Furnace and Electric Resistance Sizing

At the winter design condition, the output capacity of fossil fuel warm air furnace should not be less than the calculated total heating load. Nor should it exceed (up to a maximum of 100 percent) the design heating load by more than the next larger size that is available from the manufacturers standard product line.

At the winter design condition, the output capacity of electric resistance heating systems should not be less than the calculated total heating load, nor should it exceed the calculated load by more than 10 percent.

7-28 Cooling Equipment Sizing

Every effort should be made to obtain and use the manufacturers cooling performance date (application ratings) for selecting equipment. This published information should provide both the sensible and latent cooling capacity for the following conditions:

- Air source condensing unit operating at the summer design temperature, or water cooled condensing unit operating at the design water temperature.

- Indoor fan operating at the design CFM.

- Indoor coil operating at entering dry bulb and entering wet bulb temperatures which can be expected to occur at the summer design condition.

"Cooling only" equipment should be selected so that its sensible capacity is not less than calculated total sensible load or not more than 115 percent of this calculated load (allowing for the standard steps in capacity provided by the manufacturers product line). In addition, the corresponding latent capacity of the equipment should not be less than the calculated latent load.

In order to take full advantage of the heat pump during the heating season, heating and cooling heat pumps can be oversized for cooling by up to 25 percent if this practice produces a lower balance point, increased heating efficiency load and noticeable reduction in annual operating cost.

Multi-zone cooling equipment sizing requires a separate cooling load calculation for each zone. This calculation should include an allowance for the peak zone (sensible) load and an estimate of the zone latent load. When simultaneous "whole house" cooling is not required, multi-zone systems can be selected to provide a total cooling capacity that is less than the total calculated load for the "entire" house. Refer to Appendix 2 for more information on calculating residential zone loads and sizing multi-zone equipment.

7-29 Heat Pump Sizing

Heat pumps which are used for "heating only" applications should not be sized for less than 75 percent or more than 115 percent of the calculated heating load. Auxiliary heat should be sized to make up for any deficiency in output when the heat pump unit cannot provide full heating at the design condition.

Heat pumps which provide heating and cooling shall be sized to provide the lowest possible balance point (when operating in the heating mode) without exceeding the limits of the cooling requirements specified in Section 7-27 above. Auxiliary heat should be sized to make up for any deficiency in output when the heat pump unit cannot provide full heating at the design condition.

Multi-zone heat pump equipment sizing requires a separate heating and load calculation for each zone. Refer to Appendix 2 for more information on calculating residential zone heating loads and sizing multi-zone equipment.

Auxiliary (electrical resistance) heat should be sized to make up for the difference between the design heating load and the heat pump output on a design day. The combined output of the heat pump and the auxiliary heat should not exceed 115 percent of the calculated heating load.

Emergency or stand-by (electrical resistance) heat should be sized according to local codes or utility regulations. When emergency heat is installed to satisfy owner or contractor preferences, the emergency heat should be sized to provide 80 percent of the design heating requirement. In all cases, the emergency heat should be controlled to operate independently of the primary heating system so that it cannot be energized when the controls call for auxiliary heat. Auxiliary heat should be controlled so it can be used to provide a portion of the required emergency heat. The total installed capacity of the auxiliary heat and the emergency heat should not exceed the emergency heat requirement.

Table 1
OUTDOOR DESIGN CONDITIONS FOR UNITED STATES AND CANADA
DESIGN GRAINS BASED ON AN INSIDE DESIGN TEMPERATURE OF 75°F

Location	Latitude Degrees	WINTER 97½% Design db	WINTER Heating D.D. Below 65°F	SUMMER 2½% Design db	SUMMER Coincident Design wb	SUMMER Grains Difference 55% RH	SUMMER Grains Difference 50% RH	Daily Range	
ALABAMA									
Alexander City	33	22	....	93	76	37	44	21	M
Anniston AP	33	22	2810	94	76	35	42	21	M
Auburn	32	22	....	93	76	37	44	21	M
Birmingham AP	33	21	2710	94	75	30	37	21	M
Decatur	34	16	3050	93	74	25	32	22	M
Dothan AP	31	27	1400	92	76	39	46	20	M
Florence AP	34	21	3199	94	74	23	30	22	M
Gadsden	34	20	3000	94	75	30	37	22	M
Huntsville AP	34	16	3190	93	74	25	33	23	M
Mobile AP	30	29	1620	93	77	44	51	18	M
Mobile CO	30	29	1620	93	77	44	51	16	M
Montgomery AP	32	25	2250	95	76	33	40	21	M
Selma-Craig AFB	32	26	2160	95	77	38	47	21	M
Talladega	33	22	....	94	76	33	42	21	M
Tuscaloosa AP	33	23	2590	96	76	32	39	22	M
ALASKA									
Anchorage AP	61	-18	10860	68	58	0	0	15	L
Barrow (S)	71	-41	20265	53	50	0	0	12	L
Fairbanks AP (S)	64	-47	14290	78	60	0	0	24	M
Juneau AP	58	1	9080	70	58	0	0	15	L
Kodiak	57	13	8860	65	56	0	0	10	L
Nome AP	64	-27	14170	62	55	0	0	10	L
ARIZONA									
Douglas AP	31	31	2630	95	63	0	0	31	H
Flagstaff AP	35	4	7290	82	55	0	0	31	H
Fort Huachuca AP (S)	31	28	2551	92	62	0	0	27	H
Kingman AP	35	25	....	100	64	0	0	30	H
Nogales	31	32	2150	96	64	0	0	31	H
Phoenix AP (S)	33	34	1680	107	71	0	0	27	H
Prescott AP	34	9	....	94	60	0	0	30	H
Tuscon AP (S)	32	32	1700	102	66	0	0	26	H
Winslow AP	35	10	4780	95	60	0	0	32	H
Yuma AP	32	39	970	109	72	0	0	27	H
ARKANSAS									
Blytheville AFB	36	15	3760	94	77	42	49	21	M
Camden	33	23	....	96	76	32	39	21	M
El Dorado AP	33	23	2300	96	76	32	39	21	M
Fayetteville AP	36	12	3840	94	73	18	25	23	M
Fort Smith AP	35	17	3290	98	76	29	36	24	M
Hot Springs	34	23	2729	97	77	37	44	22	M
Jonesboro	35	15	....	94	77	42	49	21	M
Little Rock AP (S)	34	20	3170	96	77	39	46	22	M
Pine Bluff AP	34	22	2588	97	77	37	44	22	M
Texarkana AP	33	23	2530	96	77	39	46	21	M
CALIFORNIA									
Bakersfield AP	35	32	2150	101	69	0	0	32	H
Barstow	34	29	2203	104	68	0	0	37	H
Blythe AP	33	33	....	110	71	0	0	28	H
Burbank AP	34	39	1700	91	68	0	2	25	M
Chico	39	30	2835	101	68	0	0	36	H
Concord	38	27	3035	97	68	0	0	32	H
Covina	34	35	....	95	68	0	0	31	H
Crescent City AP	41	33	....	65	59	0	0	18	M
Downey	34	40	....	89	70	9	16	22	M
El Cajon	32	44	....	80	69	17	24	30	H
El Centro AP (S)	32	38	925	110	74	0	4	34	H
Escondido	33	41	....	85	68	4	11	30	H
Eureaka/Arcata AP	41	33	4640	65	59	0	0	11	L
Fairfield-Travis AFB	38	32	2725	95	67	0	0	34	H
Fresno AP (S)	36	30	2610	100	69	0	0	34	H
Hamilton AFB	38	32	3311	84	66	0	3	28	H
Laguna Beach	33	43	....	80	68	13	20	18	M
Livermore	37	27	3035	97	68	0	0	24	M
Lompoc, Vandenburg AFB	34	38	3451	70	61	0	0	20	M
Long Beach AP	33	43	1803	80	68	13	20	22	M
Los Angeles AP (S)	34	43	1960	80	68	13	20	15	L
Los Angeles CO (S)	34	40	1960	89	70	9	16	20	M
Merced-Castle AFB	37	31	2470	99	69	0	0	36	H
Modesto	37	30	....	98	68	0	0	36	H
Monterey	36	38	2750	71	61	0	0	20	M

AP - Airport
CO - City Office
S - Solar Data Available

51

Table 1 (CONTINUED)

Location	Latitude Degrees	WINTER		SUMMER				Daily Range
		97½% Design db	Heating D.D. Below 65°F	2½% Design db	Coincident Design wb	Grains Difference 55% RH	Grains Difference 50% RH	
Napa	38	32	....	96	68	0	0	30 H
Needles AP	34	33	....	110	71	0	0	27 H
Oakland AP	37	36	2940	80	63	0	0	19 M
Oceanside	33	43	....	80	68	13	20	13 L
Ontario	34	33	2009	99	69	0	5	36 H
Oxnard	34	36	....	80	64	0	0	19 M
Palmdale AP	34	22	2929	101	65	0	0	35 H
Palm Springs	33	35	....	110	70	0	0	35 H
Pasadena	34	35	1694	95	68	0	0	29 H
Petaluma	38	29	....	90	66	0	0	31 H
Pomona CO	34	30	2166	99	69	0	0	36 H
Redding AP	40	31	....	102	67	0	0	32 H
Redlands	34	33	....	99	69	0	0	33 H
Richmond	38	36	....	80	63	0	0	17 M
Riverside-March AFB (S)	33	32	2162	98	68	0	0	37 H
Sacramento AP	38	32	2700	98	70	0	2	36 H
Salinas AP	36	32	....	70	60	0	0	24 M
San Bernadino, Norton AFB	34	33	1978	99	69	0		38 H
San Diego AP	32	44	1500	80	69	17	24	12 L
San Fernando	34	39	....	91	68	0	2	38 H
San Francisco AP	37	38	3040	77	63	0	0	20 M
San Francisco CO	37	40	3040	71	62	0	3	14 L
San Jose AP	37	36	2416	81	65	0	3	26 H
San Luis Obispo	35	35	2472	88	70	10	17	26 H
Santa Ana AP	33	39	1675	85	68	4	11	28 H
Santa Barbara MAP	34	36	2470	77	66	7	14	24 H
Santa Cruz	37	38	....	71	61	0	0	28 H
Santa Maria AP (S)	34	33	2930	76	63	0	2	23 M
Santa Monica CO	34	43	....	80	68	13	20	16 M
Santa Paula	34	35	....	86	67	0	4	36 H
Santa Rosa	38	29	3065	95	67	0	0	34 H
Stockton AP	37	30	2806	97	68	0	0	37 H
Ukiah	39	29	....	95	68	0	0	40 H
Visalia	36	30	....	100	69	0	0	38 H
Yreka	41	17	....	92	64	0	0	38 H
Yuba City	39	31	....	101	67	0	0	36 H
COLORADO								
Alamosa AP	37	-6	8529	82	57	0	0	35 H
Boulder	40	0	....	91	59	0	0	27 H
Colorado Springs AP	38	2	6410	88	57	0	0	30 H
Denver AP	39	1	6150	91	59	0	0	28 H
Durango	37	-1	....	87	59	0	0	30 H
Fort Collins	40	1	....	91	59	0	0	28 H
Grand Junction AP (S)	39	7	5660	94	59	0	0	29 H
Greeley	40	4	....	94	60	0	0	29 H
LaJunta AP	38	3	5132	98	68	0	0	31 H
Leadville	39	-14	....	81	51	0	0	30 H
Pueblo AP	38	0	5480	95	61	0	0	31 H
Sterling	40	-2	....	93	62	0	0	30 H
Trinidad AP	37	3	5330	91	61	0	0	32 H
CONNECTICUT								
Bridgeport AP	41	9	5617	84	71	22	29	18 M
Hartford, Brainard Field	41	7	6170	88	73	28	35	22 M
New Haven AP	41	7	5890	84	73	33	40	17 M
New London	41	9	....	85	72	26	33	16 M
Norwalk	41	9	....	84	71	22	29	19 M
Norwich	41	7	....	86	73	31	38	18 M
Waterbury	41	2	6672	85	71	20	27	21 M
Windsor Locks, Bradley Field (S)	42	4	6350	88	72	22	29	22 M
DELAWARE								
Dover AFB	39	15	4700	90	75	36	43	18 M
Wilmington AP	39	14	4930	89	74	41	38	20 M
DISTRICT OF COLUMBIA								
Andrews AFB	38	14	....	90	74	30	37	18 M
Washington National AP	38	17	4240	91	74	29	36	18 M
FLORIDA								
Belle Glade	26	44	....	91	76	41	48	16 M

AP - Airport
CO - City Office
S - Solar Data Available

Table 1 (CONTINUED)

Location	Latitude Degrees	WINTER		SUMMER				
		97½% Design db	Heating D.D. Below 65°F	2½% Design db	Coincident Design wb	Grains Difference 55% RH	Grains Difference 50% RH	Daily Range
Cape Kennedy AP	28	38	711	88	78	59	66	15 L
Daytona Beach AP	29	35	879	90	77	49	56	15 L
Fort Lauderdale	26	46	244	91	78	53	60	15 L
Fort Myers AP	26	44	442	92	78	51	58	18 M
Fort Pierce	27	42	....	90	78	55	62	15 L
Gainesville AP (S)	29	31	1081	93	77	45	51	18 M
Jacksonville AP	30	32	1230	94	77	42	49	19 M
Key West AP	24	57	110	90	78	55	62	9 L
Lakeland CO (S)	28	41	661	91	76	51	48	17 M
Miami AP (S)	25	47	200	90	77	49	56	15 L
Miami Beach CO	25	48	200	89	77	51	58	10 L
Ocala	29	34	....	93	77	44	51	18 M
Orlando AP	28	38	720	93	76	37	44	17 M
Panama City, Tyndall AFB	30	33	1390	90	77	49	56	14 L
Pensacola CO	30	29	1480	93	77	44	51	14 L
St. Augustine	29	35	1051	89	78	57	64	16 M
St. Petersburg	28	40	670	91	77	47	54	16 M
Sanford	28	38	....	93	76	37	44	17 M
Sarasota	27	42	....	92	77	45	52	17 M
Tallahassee AP (S)	30	30	1520	92	76	39	46	19 M
Tampa AP (S)	28	40	700	91	77	47	54	17 M
West Palm Beach AP	26	45	270	91	78	53	60	16 M
GEORGIA								
Albany, Turner AFB	31	29	1760	95	76	33	40	20 M
Americus	32	25	....	94	76	35	42	20 M
Athens	34	22	2929	92	74	27	34	21 M
Atlanta AP (S)	33	22	2990	92	74	27	34	19 M
Augusta AP	33	23	2410	95	76	33	40	19 M
Brunswick	31	32	1531	89	78	57	64	18 M
Columbus, Lawson AFB	32	24	2380	93	76	37	44	21 M
Dalton	34	22	....	93	76	37	44	22 M
Dublin	32	25	....	93	76	37	44	20 M
Gainesville	34	21	....	91	74	29	36	21 M
Griffin (S)	33	22	....	90	75	36	43	21 M
La Grange	33	23	....	91	75	35	42	21 M
Macon AP	32	25	2160	93	76	37	44	22 M
Marietta, Dobbins AFB	34	21	3080	92	74	27	34	21 M
Moultrie	31	30	....	95	77	40	47	20 M
Rome AP	34	22	3326	93	76	37	44	23 M
Savannah-Travis AP	32	27	1850	93	77	44	51	20 M
Valdosta-Moody AFB	31	31	1520	94	77	43	49	20 M
Waycross	31	29	....	94	77	43	49	20 M
HAWAII								
Hilo AP (S)	19	62	0	83	72	30	37	15 L
Honolulu AP	21	63	0	86	73	31	38	12 L
Kaneohe Bay MCAS	21	66	0	84	74	40	47	12 L
Wahaiwa	21	59	10	85	72	26	33	14 L
IDAHO								
Boise AP (S)	43	10	5830	94	64	0	0	31 H
Burley	42	2	....	95	61	0	0	35 H
Coeur D'Alene AP	47	-1	6660	86	61	0	0	31 H
Idaho Falls AP	43	-6	7890	87	61	0	0	38 H
Lewiston AP	46	6	5500	93	64	0	0	32 H
Moscow	46	0	....	87	62	0	0	32 H
Mountain Home AFB	43	12	6120	97	63	0	0	36 H
Pocatello AP	43	-1	7030	91	60	0	0	35 H
Twin Falls AP (S)	42	2	6730	95	61	0	0	34 H
ILLINOIS								
Aurora	41	-1	6660	91	76	41	48	20 M
Belleville, Scott AFB	38	6	4480	92	76	39	49	21 M
Bloomington	40	-2	....	90	74	30	37	21 M
Carbondale	37	7	....	93	77	44	51	21 M
Champaign/Urbana	40	2	5800	92	74	27	34	21 M
Chicago, Midway AP	41	0	6160	91	73	23	29	20 M
Chicago, O'Hare AP	42	-4	6640	89	74	31	38	20 M
Chicago CO	41	2	6640	91	74	29	36	15 L
Danville	40	1	5538	90	74	30	37	21 M
Decatur	39	2	5480	91	74	29	36	21 M
Dixon	41	-2	....	90	74	30	37	23 M
Elgin	42	-2	....	88	74	33	40	21 M
Freeport	42	-4	....	89	73	26	33	24 M

Table 1 (CONTINUED)

Location	Latitude Degrees	WINTER		SUMMER				
		97½ % Design db	Heating D.D. Below 65°F	2½ % Design db	Coincident Design wb	Grains Difference 55% RH	Grains Difference 50% RH	Daily Range
Galesburg	41	-2	6005	91	75	35	42	22 M
Greenville	39	4	....	92	75	33	40	21 M
Joliet	41	0	6180	90	74	30	37	20 M
Kankakee	41	1	6040	90	74	30	37	21 M
LaSalle/Peru	41	-2	....	91	75	35	42	22 M
Macomb	40	0	....	92	76	39	46	22 M
Moline AP	41	-4	6410	91	75	35	42	23 M
Mt. Vernon	38	5	....	92	75	33	40	21 M
Peoria AP	40	-4	6070	89	74	31	38	22 M
Quincy AP	40	3	5267	93	76	37	44	22 M
Rantoul, Chanute AFB	40	1	5966	91	74	29	36	21 M
Rockford	42	-4	6840	89	73	26	33	24 M
Springfield AP	39	2	5530	92	74	27	34	21 M
Waukegan	42	-3	....	89	74	31	37	21 M
INDIANA								
Anderson	40	6	5580	92	75	33	40	22 M
Bedford	38	5	....	92	75	33	40	22 M
Bloomington	39	5	4860	92	75	33	40	22 M
Columbus, Bakalar AFB	39	7	5132	92	75	33	40	22 M
Crawfordsville	40	3	....	91	74	29	36	22 M
Evansville AP	38	9	4500	93	75	31	38	22 M
Fort Wayne AP	41	1	6220	89	72	20	27	24 M
Goshen AP	41	1	....	89	73	26	33	23 M
Hobart	41	2	....	88	73	26	33	21 M
Huntington	40	1	....	89	72	20	27	23 M
Indianapolis AP (S)	39	2	5630	90	74	30	37	22 M
Jeffersonville	38	10	....	93	74	25	33	23 M
Kokoma	40	0	5590	90	73	24	31	22 M
Layfayette	40	3	5820	91	73	23	30	22 M
LaPorte	41	3	....	90	74	30	37	22 M
Marion	40	0	....	90	73	24	31	23 M
Muncie	40	2	5950	90	73	24	31	22 M
Peru, Bunker Hill AFB	40	-1	....	88	73	28	35	22 M
Richmond AP	39	2	....	90	74	30	37	22 M
Shelbyville	39	3	....	91	74	29	36	22 M
South Bend AP	41	1	6460	89	73	26	33	22 M
Terre Haute AP	39	4	5360	92	74	27	34	22 M
Valparaise	41	3	....	90	74	30	37	22 M
Vincennes	38	6	....	92	74	27	34	22 M
IOWA								
Ames (S)	42	-6	....	90	74	30	37	23 M
Burlington AP	40	-3	6120	91	75	35	43	22 M
Cedar Rapids AP	41	-5	6600	88	75	38	45	23 M
Clinton	41	-3	....	90	75	36	33	23 M
Council Biuffs	41	-3	6610	91	75	35	42	22 M
Des Moines AP	41	-5	6610	91	74	29	36	23 M
Debuque	42	-7	7380	88	73	28	35	22 M
Fort Dodge	42	-7	7070	88	74	33	40	23 M
Iowa City	41	-6	6404	89	76	44	51	22 M
Keokuk	40	0	6404	92	75	33	40	22 M
Marshalltown	42	-7	6850	90	75	36	43	23 M
Mason City AP	43	-11	7790	88	74	33	40	24 M
Newton	41	-5	....	91	74	29	36	23 M
Ottumwa AP	41	-4	....	91	74	29	36	22 M
Sioux City AP	42	-7	6960	92	74	27	34	24 M
Waterloo	42	-10	7370	89	75	38	45	23 M
KANSAS								
Atchison	39	2	....	93	76	37	44	23 M
Chanute AP	34	7	4566	97	74	19	26	23 M
Dodge City AP (S)	37	5	....	97	69	0	0	25 M
El Dorado	37	7	4990	98	73	11	18	24 M
Emporia	38	5	....	97	74	19	26	25 M
Garden City AP	38	4	....	96	69	0	0	28 M
Goodland AP	39	0	6140	96	65	0	0	31 H
Great Bend	38	4	....	98	73	11	18	28 H
Hutchinson AP	38	8	4680	99	72	3	10	28 H
Liberal	37	7	....	96	68	0	0	28 H
Manhattan, Fort Riley (S)	39	3	5306	95	75	28	35	24 M
Parsons	37	9	4158	97	74	19	26	23 M
Russel AP	38	4	....	98	73	11	18	29 H
Salina	38	5	4970	100	74	13	20	26 M
Topeka AP	39	4	5210	96	75	26	33	24 M

AP - Airport
CO - City Office
S - Solar Data Available

Table 1 (CONTINUED)

Location	Latitude Degrees	WINTER 97½% Design db	WINTER Heating D.D. Below 65°F	SUMMER 2½% Design db	Coincident Design wb	Grains Difference 55% RH	Grains Difference 50% RH	Daily Range
Wichita AP	37	7	4640	98	73	11	18	23 M
KENTUCKY								
Ashland	38	10	4555	91	74	29	36	22 M
Bowling Green AP	37	10	4280	92	75	33	40	21 M
Corbin AP	37	9	....	92	73	21	28	23 M
Covington AP	39	6	5260	90	72	19	26	22 M
Hopkinsville, Campbell AFB	36	10	4290	92	75	33	40	21 M
Lexington AP (S)	38	8	4760	91	73	23	30	22 M
Louisville AP	38	10	4610	93	74	25	33	23 M
Madisonville	37	10	....	93	75	31	38	22 M
Owensboro	37	10	4200	94	75	30	37	23 M
Paducah	37	12	3650	95	75	28	35	20 M
LOUISIANA								
Alexandria AP	31	27	2000	94	77	43	49	20 M
Baton Rouge AP	30	29	1610	93	77	44	51	19 M
Bogalusa	30	28	....	93	77	44	51	19 M
Houma	29	35	....	93	78	50	57	15 L
Lafayette AP	30	30	1550	94	78	49	56	18 M
Lake Charles AP (S)	30	31	1490	93	77	44	51	17 M
Minden	32	25	....	96	76	32	38	20 M
Monroe AP	32	25	2310	96	76	32	38	20 M
Natchitoches	31	26	....	95	77	40	47	20 M
New Orleans AP	30	33	1400	92	78	51	58	16 M
Shreveport AP (S)	32	25	2160	96	76	32	39	20 M
MAINE								
Augusta AP	44	-3	7826	85	70	15	23	22 M
Bangor, Dow AFB	44	-6	8220	83	68	8	15	22 M
Caribou AP (S)	46	-13	9770	81	67	7	14	21 M
Lewiston	44	-2	7690	85	70	15	22	22 M
Millinocket AP	45	-9	8533	83	68	8	15	22 M
Portland (S)	43	-1	7570	84	71	23	29	22 M
Waterville	44	-4	....	84	69	11	18	22 M
MARYLAND								
Baltimore AP	39	13	4680	91	75	35	42	21 M
Baltimore CO	39	17	....	89	76	43	51	17 M
Cumberland	39	10	5070	89	74	31	38	22 M
Frederick AP	39	12	5030	91	75	35	42	22 M
Hagerstown	39	12	5130	91	74	29	36	22 M
Salisbury (S)	38	16	4220	91	75	35	42	18 M
MASSACHUSETTS								
Boston AP (S)	42	9	5630	88	71	16	23	16 M
Clinton	42	2	....	87	71	18	25	17 M
Fall River	41	9	5774	84	71	22	29	18 M
Framingham	42	6	....	86	71	19	26	17 M
Gloucester	42	5	....	86	71	19	26	15 L
Greenfield	42	-2	....	85	71	20	27	23 M
Lawrence	42	0	6195	87	72	23	30	22 M
Lowell	42	1	6060	88	72	22	29	21 M
New Bedford	41	9	5400	82	71	26	33	19 M
Pittsfield AP	42	-3	7580	84	70	16	23	23 M
Springfield, Westover AFB	42	0	5840	87	71	18	25	19 M
Tauton	41	9	....	86	72	25	32	18 M
Worcester AP	42	4	6970	84	70	16	23	18 M
MICHIGAN								
Adrian	41	3	....	88	72	22	29	23 M
Alpena AP	45	-6	8510	85	70	15	22	27 H
Battle Creek AP	42	5	6580	88	72	22	29	23 M
Benton Harbor AP	42	5	....	88	72	22	29	20 M
Detroit	42	6	6290	88	72	22	29	20 M
Escanaba	45	-7	8481	83	69	13	20	17 M
Flint AP	42	1	7200	87	72	23	30	25 M
Grand Rapids AP	42	5	6890	88	72	22	29	24 M
Holland	42	6	....	86	71	19	26	22 M
Jackson AP	42	5	....	88	72	22	29	23 M
Kalamazoo	42	5	....	88	72	22	29	23 M
Lansing AP	42	1	6940	87	72	23	30	24 M
Marquette CO	46	-8	8390	81	69	16	23	18 M

AP - Airport
CO - City Office
S - Solar Data Available

Table 1 (CONTINUED)

Location	Latitude Degrees	WINTER 97½% Design db	WINTER Heating D.D. Below 65°F	SUMMER 2½% Design db	SUMMER Coincident Design wb	SUMMER Grains Difference 55% RH	SUMMER Grains Difference 50% RH	Daily Range	
Mt. Pleasant	43	4	....	87	72	23	30	24	M
Muskegon AP	43	6	6700	84	70	16	23	21	H
Pontiac	42	4	....	87	72	23	30	21	M
Port Huron	43	4	6564	87	72	23	30	21	H
Saginaw AP	43	4	7120	87	72	23	30	23	M
Sault Ste. Marie AP (S)	46	-8	9050	81	69	16	23	23	M
Traverse City AP	44	1	7700	86	71	19	26	22	M
Yipsilanti	42	5	6424	89	71	15	22	22	M
MINNESOTA									
Albert Lea	43	-12	....	87	72	23	30	24	M
Alexandria AP	45	-16	....	88	72	22	29	24	M
Bemidji AP	47	-26	10203	85	69	9	16	24	M
Brainerd	46	-16	....	87	71	18	25	24	M
Duluth AP	46	-16	9890	82	68	10	17	22	M
Fairbault	44	-12	....	88	72	22	27	24	M
Fergus Falls	46	-17	....	88	72	22	27	24	M
International Falls AP	48	-25	10600	83	68	8	15	26	M
Mankato	44	-12	8310	88	72	22	29	24	M
Minneapolis/St. Paul AP	44	-12	8250	89	73	26	33	22	M
Rochester AP	44	-12	8295	87	72	23	30	24	M
St. Cloud AP (S)	45	-11	8890	88	72	22	29	24	M
Virginia	47	-21	....	83	68	8	15	23	M
Wilmar	45	-11	....	88	72	22	29	24	M
Winona	44	-10	....	88	73	28	35	24	M
MISSISSIPPI									
Biloxi, Keesler AFB	30	31	1500	92	79	58	65	16	M
Clarksdale	34	19	....	94	77	42	49	21	M
Columbus AFB	33	20	2890	93	77	44	51	22	M
Greenville AFB	33	20	2580	93	77	44	51	21	M
Greenwood	33	20	....	93	77	44	51	21	M
Hattiesburg	31	27	1840	94	77	42	49	21	M
Jackson AP	32	25	2260	95	76	33	40	21	M
Laurel	31	27	....	94	77	42	49	21	M
McComb AP	31	26	....	94	76	35	42	18	M
Meridian AP	32	23	2340	95	76	33	40	22	M
Natchez	31	27	1800	94	78	49	56	21	M
Tupelo	34	19	....	94	77	42	49	22	M
Vicksburg CO	32	26	2040	95	78	47	54	21	M
MISSOURI									
Cape Girardeau	37	13	....	95	75	28	35	21	M
Columbia AP (S)	39	4	5070	94	74	23	30	22	M
Farmington AP	37	8	....	93	75	31	38	22	M
Hannibal	39	3	5512	93	76	37	44	22	M
Jefferson City	38	7	4620	95	74	21	28	23	M
Joplin AP	37	10	4090	97	73	13	20	24	M
Kansas City AP	39	6	4750	96	74	20	27	20	M
Kirksville AP	40	0	....	93	74	25	32	24	M
Mexico	39	4	....	94	74	23	30	22	M
Moberly	39	3	....	94	74	23	30	23	M
Poplar Bluff	36	16	3910	95	76	32	40	22	M
Rolla	38	9	....	91	75	35	42	22	M
St. Joseph AP	39	2	5440	93	76	37	44	23	M
St. Louis AP	38	6	4900	94	75	30	37	21	M
St. Louis CO	38	8	....	94	75	30	37	18	M
Sedalia, Whiteman AFB	38	4	5012	92	76	39	46	22	M
Sikeston	36	15	....	95	76	33	40	21	M
Springfield AP	37	9	4900	93	74	25	32	23	M
MONTANA									
Billings AP	45	-10	7150	91	64	0	0	31	H
Bozeman	45	-14	....	87	60	0	0	32	H
Butte AP	46	-17	9730	83	56	0	0	35	H
Cut Bank AP	48	-20	9033	85	61	0	0	35	H
Glasgow AP (S)	48	-18	9000	89	63	0	0	29	H
Glendive	47	-13	....	92	64	0	0	29	H
Great Falls AP (S)	47	-15	7670	88	60	0	0	28	H
Havre	48	-11	8880	90	64	0	0	33	H
Helena AP	46	-16	8190	88	60	0	0	32	H
Kalispell AP	48	-7	8150	87	61	0	0	34	H
Lewiston AP	47	-16	8586	87	61	0	0	30	H
Livingston AP	45	-14	....	87	60	0	0	32	H

AP - Airport
CO - City Office
S - Solar Data Available

Table 1 (CONTINUED)

Location	Latitude Degrees	WINTER 97½% Design db	WINTER Heating D.D. Below 65°F	SUMMER 2½% Design db	SUMMER Coincident Design wb	SUMMER Grains Difference 55% RH	SUMMER Grains Difference 50% RH	Daily Range
Miles City AP	46	15	7810	95	66	0	0	30 H
Missoula AP	46	-6	8000	88	61	0	0	36 H
NEBRASKA								
Beatrice	40	-2	....	95	74	21	28	24 M
Chadron AP	42	-3	7100	94	65	0	0	30 H
Columbus	41	-2	....	95	73	16	23	M
Fremont	41	-2	....	95	74	21	28	22 M
Grand Island AP	41	-3	6440	94	71	6	13	28 H
Hastings	40	-3	6070	94	71	6	13	27 H
Kearney	40	-4	....	93	70	2	9	28 H
Lincoln CO (S)	40	-2	6050	95	74	21	28	24 M
McCook	40	-2	....	95	69	0	0	28 H
Norfolk	42	-4	7010	93	74	25	32	30 H
North Platte AP (S)	41	-4	6680	94	69	0	2	28 H
Omaha AP	41	-3	6290	91	75	35	42	22 M
Scottsbluff AP	41	-3	6670	92	65	0	0	31 H
Sidney AP	41	-3	7030	92	65	0	0	31 H
NEVADA								
Carson City	39	9	5753	91	59	0	0	42 H
Elko AP	40	-2	7430	92	59	0	0	42 H
Ely AP (S)	39	-4	7710	87	56	0	0	39 H
Las Vegas AP (S)	36	28	2610	106	65	0	0	30 H
Lovelock AP	40	12	....	96	63	0	0	42 H
Reno AP (S)	39	10	6150	92	60	0	0	45 H
Reno CO	39	11	....	93	60	0	0	45 H
Tonapah AP	38	10	5900	92	59	0	0	40 H
Winnemucca AP	40	3	6760	94	60	0	0	42 H
NEW HAMPSHIRE								
Berlin	44	-9	8270	84	69	11	18	22 M
Claremont	43	-4	7850	86	70	14	21	24 M
Concord AP	43	-3	7360	87	70	12	19	26 H
Keene	43	-7	7460	87	70	12	19	24 M
Laconia	43	-5	7560	86	70	14	21	25 M
Manchester, Grenier AFB	43	-3	7100	88	71	16	23	24 M
Portsmouth, Pease AFB	43	2	6710	85	71	20	27	22 M
NEW JERSEY								
Atlantic City CO	39	13	4810	89	74	31	38	18 M
Long Branch	40	13	....	90	73	24	31	18 M
Newark AP	40	14	4900	91	73	23	30	20 M
New Brunswick	40	10	5400	89	73	26	33	19 M
Paterson	40	10	5360	91	73	23	30	21 M
Phillipsburg	40	6	....	89	72	20	27	21 M
Trenton CO	40	14	4980	88	74	33	40	19 M
Vineland	39	11	....	89	74	31	38	19 M
NEW MEXICO								
Alamagordo Holloman AFB	32	19	3240	96	64	0	0	30 H
Albuquerque AP (S)	35	16	4250	94	61	0	0	27 H
Artesia	32	19	....	100	67	0	0	30 H
Carlsbad AP	32	19	2835	100	67	0	0	28 H
Clovis AP	34	13	4200	93	65	0	0	28 H
Farmington AP	36	6	5713	93	62	0	0	30 H
Gallup	35	5	....	89	58	0	0	32 H
Grants	35	4	....	88	58	0	0	32 H
Hobbs AP	32	18	....	99	66	0	0	29 H
Las Cruces	32	20	3194	96	64	0	0	30 H
Los Alamos	35	9	....	87	60	0	0	32 H
Raton AP	36	1	6228	89	60	0	0	34 H
Roswell, Walker AFB	33	18	3680	98	66	0	0	33 H
Santa Fe CO	35	10	6120	88	61	0	0	28 H
Silver City AP	32	10	3705	94	60	0	0	30 H
Socorro AP	34	17	....	95	62	0	0	30 H
Tucumcari AP	35	13	4047	97	66	0	0	28 H
NEW YORK								
Albany AP (S)	42	-1	6900	88	72	22	29	23 M
Albany CO	42	1	....	88	72	22	29	20 M
Auburn	43	2	....	87	71	18	25	22 M

AP - Airport
CO - City Office
S - Solar Data Available

Table 1 (CONTINUED)

Location	Latitude Degrees	WINTER		SUMMER				
		97½% Design db	Heating D.D. Below 65°F	2½% Design db	Coincident Design wb	Grains Difference 55% RH	Grains Difference 50% RH	Daily Range
Batavia	43	5	....	87	71	18	25	22 M
Binghamton AP	42	1	7340	83	69	13	20	20 M
Buffalo AP	43	6	6960	85	70	15	22	21 M
Cortland	42	0	....	85	71	20	27	23 M
Dunkirk	42	9	6851	85	72	26	33	18 M
Elmira AP	42	1	....	86	71	19	26	24 M
Geneva (S)	42	2	....	87	71	18	20	22 M
Glen Falls	42	-5	7270	85	71	20	27	23 M
Gloversville	43	-2	....	86	71	19	26	23 M
Hornell	42	0	....	85	70	15	22	24 M
Ithaca (S)	42	0	7052	85	71	20	27	24 M
Jamestown	42	3	6849	86	70	14	21	20 M
Kingston	42	2	....	88	72	22	29	22 M
Lockport	43	7	6724	86	72	25	32	21 M
Massena AP	45	-8	....	83	69	13	20	20 M
Newburg-Stewart AFB	41	4	6336	88	72	22	29	21 M
NYC-Central Park (S)	40	15	4880	89	73	26	33	17 M
NYC-Kennedy AP	40	15	5219	87	72	23	30	16 M
NYC-La Guardia AP	40	15	4811	89	73	26	33	16 M
Niagra Falls AP	43	7	6688	86	72	25	32	20 M
Olean	42	2	....	84	71	22	29	23 M
Oneonta	42	-4	....	83	69	13	20	24 M
Oswego CO	43	7	6792	83	71	24	31	20 M
Plattsburg AFB	44	-8	8044	83	69	13	20	22 M
Poughkeepsie	41	6	5820	89	74	31	38	21 M
Rochester AP	43	5	6760	88	71	16	23	22 M
Rome-Griffiss AFB	43	-5	7331	85	70	15	22	22 M
Schenectady (S)	42	1	6650	87	72	23	30	22 M
Suffolk County AFB	40	10	5951	83	71	24	31	16 M
Syracuse AP	43	2	6720	87	71	18	25	20 M
Utica	43	-6	7200	85	71	20	27	22 M
Watertown	44	-6	7300	83	71	24	31	20 M
NORTH CAROLINA								
Ashville AP	35	14	4130	87	72	23	30	21 M
Charlotte AP	35	22	3200	93	74	25	32	20 M
Durham	36	20	....	92	75	33	40	20 M
Elizabeth City AP	36	19	3207	91	77	47	54	18 M
Fayetteville, Pope AFB	35	20	3080	92	76	39	46	20 M
Goldsboro, Seymour-Johnson AFB	35	21	3124	91	76	41	48	18 M
Greensboro AP (S)	36	18	3810	91	73	23	30	21 M
Greenville	35	21	....	91	76	41	48	19 M
Henderson	36	15	....	92	76	39	46	20 M
Hickory	35	18	....	90	72	19	26	21 M
Jacksonville	34	24	....	90	78	55	62	18 M
Lumberton	34	21	....	92	76	39	46	20 M
New Bern AP	35	24	....	90	78	55	62	18 M
Raleigh/Durham AP (S)	35	20	3440	92	75	33	40	20 M
Rocky Mount	36	21	....	91	76	41	48	19 M
Wilmington AP	34	26	2380	91	78	53	60	18 M
Winston-Salem AP	36	20	3650	91	73	23	30	20 M
NORTH DAKOTA								
Bismark AP (S)	46	-19	8960	91	68	0	2	27 H
Devil's Lake	48	-21	9901	88	68	0	7	25 M
Dickinson AP	46	-17	8942	90	66	0	0	25 M
Fargo AP	46	-18	9250	89	71	15	22	25 M
Grands Forks AP	48	-22	9930	87	70	12	19	25 M
Jamestown AP	47	-18	....	90	69	2	9	26 H
Minot AP	48	-20	9610	89	67	0	0	25 M
Williston	48	-21	9180	88	67	0	1	25 M
OHIO								
Akron-Canton AP	41	6	6140	86	71	19	26	21 M
Ashtabula	41	9	....	85	72	26	33	18 M
Athens	39	6	....	92	74	27	34	22 M
Bowling Green	41	2	....	89	73	26	33	23 M
Cambridge	40	7	....	90	74	30	37	23 M
Chilicothe	39	6	....	92	74	27	34	22 M
Cincinnati CO	39	6	4830	90	72	19	26	21 M
Cleveland AP (S)	41	5	6200	88	72	22	29	22 M
Columbus AP (S)	40	5	5670	90	73	24	31	24 M
Dayton AP	39	4	5620	89	72	20	27	20 M
Defiance	41	4	....	91	73	23	30	24 M

AP - Airport
CO - City Office
S - Solar Data Available

Table 1 (CONTINUED)

Location	Latitude Degrees	WINTER		SUMMER				
		97½% Design db	Heating D.D. Below 65°F	2½% Design db	Coincident Design wb	Grains Difference 55% RH	Grains Difference 50% RH	Daily Range
Findlay AP	41	3	....	90	73	24	31	24 M
Fremont	41	1	....	88	73	28	35	24 M
Hamilton	39	5	....	90	72	19	26	22 M
Lancaster	39	5	....	91	73	23	30	23 M
Lima	40	4	5870	91	73	23	30	24 M
Mansfield AP	40	5	6403	87	72	23	30	22 M
Marion	40	5	....	91	73	23	30	23 M
Middletown	39	5	....	90	72	19	26	22 M
Newark	40	5	5655	92	73	21	28	23 M
Norwalk	41	1	....	88	73	28	35	22 M
Portsmouth	38	10	4410	92	74	27	34	22 M
Sandusky CO	41	6	5796	91	72	17	24	21 M
Springfield	40	3	4284	89	73	26	33	21 M
Stubenville	40	5	....	86	71	19	26	22 M
Toledo AP	41	1	6430	88	73	28	35	25 M
Warren	41	5	....	87	71	18	25	23 M
Wooster	40	6	....	86	71	19	26	22 M
Youngstown AP	41	4	6370	86	71	19	26	23 M
Zanesville AP	40	7	....	90	74	30	37	23 M
OKLAHOMA								
Ada	34	14	....	97	74	19	26	23 M
Altus AFB	34	16	3190	100	73	7	14	25 M
Ardmore	34	17	3060	98	74	17	24	23 M
Bartlesville	36	10	....	98	74	17	24	23 M
Chickasha	35	14	....	98	74	17	24	24 M
Enid-Vance AFB	36	13	3971	100	74	13	20	24 M
Lawton AP	34	16	3250	99	74	15	22	24 M
McAlester	34	19	3255	96	74	20	27	23 M
Muskogee AP	35	15	....	98	75	23	30	23 M
Norman	35	13	3247	96	74	20	27	24 M
Oklahoma City AP	35	13	3700	97	74	19	26	23 M
Ponca City	36	9	3850	97	74	19	26	24 M
Seminole	35	15	....	96	74	20	27	23 M
Stillwater (S)	36	13	....	96	74	20	27	24 M
Tulsa AP	36	13	3730	98	75	23	30	22 M
Woodward	36	10	....	97	73	13	20	26 H
OREGON								
Albany	44	22	....	89	66	0	0	31 H
Astoria AP (S)	46	29	5190	71	62	0	4	16 M
Baker AP	44	6	....	89	61	0	0	30 H
Bend	44	4	....	87	60	0	0	33 H
Corvallis (S)	44	22	4854	89	66	0	0	31 H
Eugene AP	44	22	4740	89	66	0	0	31 H
Grants Pass	42	24	4375	96	68	0	0	33 H
Klamath Falls AP	42	9	6810	87	60	0	0	36 H
Medford AP (S)	42	23	4880	94	67	0	0	35 H
Pendleton AP	45	5	4700	93	64	0	0	29 H
Portland AP	45	23	4635	85	67	0	7	23 M
Portland CO	45	24	....	86	67	0	4	21 M
Roseburg AP	43	23	4491	90	66	0	0	30 H
Salem AP	45	23	4760	88	66	0	0	31 H
The Dalles	45	19	....	89	68	3	10	28 H
PENNSYLVANIA								
Allentown AP	40	9	5810	88	72	22	29	22 M
Altoona CO	40	5	6192	87	71	18	25	23 M
Butler	40	6	....	87	72	23	30	22 M
Chambersburg	40	8	5170	90	74	30	37	23 M
Erie AP	42	9	6540	85	72	26	33	18 M
Harrisburg AP	40	11	5280	91	74	29	36	21 M
Johnstown	40	2	7804	83	70	19	26	23 M
Lancaster	40	8	5560	90	74	30	37	22 M
Meadville	41	4	....	85	70	15	32	21 M
New Castle	41	7	5800	88	72	22	30	23 M
Philadelphia AP	39	14	5180	90	74	30	37	21 M
Pittsburgh AP	40	5	5950	86	71	19	26	22 M
Pittsburgh CO	40	7	....	88	71	16	22	19 M
Reading CO	40	13	4960	89	72	20	22	19 M
Scranton/Wilkes-Barre	41	5	6160	87	71	18	25	19 M
State College (S)	40	7	6160	87	71	18	25	23 M
Sunbury	40	7	....	89	72	20	27	22 M
Uniontown	39	9	....	88	73	28	35	22 M
Warren	41	4	....	86	71	19	26	24 M

AP - Airport
CO - City Office
S - Solar Data Available

Table 1 (CONTINUED)

Location	Latitude Degrees	WINTER 97½% Design db	WINTER Heating D.D. Below 65°F	SUMMER 2½% Design db	SUMMER Coincident Design wb	SUMMER Grains Difference 55% RH	SUMMER Grains Difference 50% RH	Daily Range
West Chester	40	13	....	89	74	31	38	20 M
Williamsport AP	41	7	5950	89	72	20	27	23 M
York	40	12	....	91	74	29	36	22 M
RHODE ISLAND								
Newport (S)	41	9	5800	85	72	26	33	16 M
Providence AP	41	9	5950	86	72	25	32	19 M
SOUTH CAROLINA								
Anderson	34	23	....	92	74	27	34	21 M
Charleston AFB (S)	32	27	2070	91	78	53	60	18 M
Charleston CO	32	28	2146	92	78	51	58	13 L
Columbia AP	34	24	2520	95	75	28	35	22 M
Florence AP	34	25	2480	92	77	45	52	21 M
Georgetown	33	26	2228	90	78	55	62	18 M
Greenville AP	34	22	3070	91	74	29	36	21 M
Greenwood	34	22	2890	93	74	25	32	21 M
Orangeburg	33	24	....	95	75	28	35	20 M
Rock Hill	35	23	....	94	74	23	30	20 M
Spartanburg AP	35	22	....	91	74	29	36	20 M
Sumter-Shaw AFB	34	25	2453	92	76	39	46	21 M
SOUTH DAKOTA								
Aberdeen AP	45	-15	8620	91	72	17	24	27 H
Brookings	44	-13	....	92	72	15	22	25 M
Huron AP	44	-14	8220	93	72	13	20	28 H
Mitchel	43	-10	....	93	71	7	14	28 H
Pierre AP	44	-10	7550	95	71	4	11	29 H
Rapid City AP (S)	44	-7	7370	92	65	0	0	28 H
Sioux Falls AP	43	-11	7840	91	72	17	24	24 M
Watertown AP	45	-15	8390	91	72	17	24	26 H
Yankton	43	-7	....	91	72	17	24	25 M
TENNESSEE								
Athens	33	18	....	92	73	21	28	22 M
Bristol-Tri City AP	36	14	4143	89	72	20	27	22 M
Chattanooga AP	35	18	3380	93	74	25	32	22 M
Clarksville	36	12	....	93	74	25	32	21 M
Columbia	35	15	....	94	74	23	30	21 M
Dyersburg	36	15	....	94	77	42	49	21 M
Greenville	35	16	....	90	72	19	26	22 M
Jackson AP	35	16	3350	95	75	28	35	21 M
Knoxville AP	35	19	3510	92	73	21	28	21 M
Memphis AP	35	18	3210	95	76	33	48	21 M
Murfreesboro	35	14	....	94	74	23	30	22 M
Nashville AP (S)	36	14	3610	94	74	23	30	21 M
Tullahoma	35	13	3577	93	73	19	26	22 M
TEXAS								
Abilene AP	32	20	2620	99	71	0	5	22 M
Alice AP	27	34	....	98	77	35	42	20 M
Amarillo AP	35	11	4140	95	67	33	40	26 H
Austin AP	30	28	1720	98	74	23	30	22 M
Bay City	29	33	....	94	77	42	49	16 M
Beaumont	30	31	1370	93	78	50	57	19 M
Beeville	28	33	1189	97	77	37	44	18 M
Big Springs AP (S)	32	20	....	97	69	0	0	26 H
Brownsville AP (S)	25	39	600	93	77	44	51	18 M
Brownwood	31	22	2437	99	73	9	16	22 M
Bryan AP	30	29	1640	96	76	32	39	20 M
Corpus Christi AP	27	35	930	94	78	49	56	19 M
Corsicana	32	25	....	98	75	23	30	21 M
Dallas AP	32	22	2320	100	75	20	27	20 M
Del Rio, Laughlin AFB	29	31	1520	98	73	11	18	24 M
Denton	33	22	....	99	74	15	22	22 M
Eagle Pass	28	32	1423	99	73	9	16	24 M
El Paso AP (S)	31	24	2680	98	64	0	0	27 H
Fort Worth AP (S)	32	22	2390	99	74	15	22	22 M
Galveston AP	29	36	1270	89	79	63	70	10 L
Greenville	33	22	....	99	74	15	22	21 M
Harlingen	26	39	693	94	77	42	49	19 M
Houston AP	29	32	1410	94	77	42	49	18 M
Houston CO	29	33	....	95	77	40	47	18 M
Huntsville	30	27	....	98	75	23	30	20 M
Killeen-Gray AFB	31	25	....	97	73	13	20	22 M

AP - Airport
CO - City Office
S - Solar Data Available

Table 1 (CONTINUED)

Location	Latitude Degrees	WINTER 97½% Design db	WINTER Heating D.D. Below 65°F	SUMMER 2½% Design db	SUMMER Coincident Design wb	SUMMER Grains Difference 55% RH	SUMMER Grains Difference 50% RH	Daily Range
Lamesa	32	17	....	96	69	0	0	26 H
Laredo AFB	27	36	800	101	73	6	13	23 M
Longview	32	24	....	97	76	30	37	20 M
Lubbock AP	33	15	3570	96	69	0	0	26 H
Lufkin AP	31	29	1940	97	76	30	37	20 M
McAllen	26	39	....	95	77	40	47	21 M
Midland AP (S)	32	21	2600	98	69	0	0	26 H
Mineral Wells AP	32	22	....	99	74	15	22	22 M
Palestine CO	31	27	....	98	76	29	36	20 M
Pampa	35	12	....	96	67	0	0	26 H
Pecos	31	21	....	98	69	0	0	27 H
Plainview	34	13	....	96	68	0	0	26 H
Port Arthur AP	30	31	1447	93	78	50	57	19 M
San Angelo, Goodfellow AFB	31	22	2220	99	71	0	5	24 M
San Antonio AP (S)	29	30	1560	97	73	13	20	19 M
Sherman-Perrin AFB	33	20	2837	98	75	23	30	22 M
Snyder	32	18	....	98	70	0	1	26 M
Temple	31	27	....	99	74	15	22	22 M
Tyler AP	32	24	....	97	76	30	37	21 M
Vernon	34	17	....	100	73	7	14	24 M
Victoria AP	28	32	1160	96	77	39	46	18 M
Waco AP	31	26	2040	99	75	21	28	22 M
Wichita Falls AP	34	18	2900	101	73	6	13	24 M
UTAH								
Cedar City AP	37	5	5680	91	60	0	0	32 H
Logan	41	2	6750	91	61	0	0	33 H
Moab	38	11	....	98	60	0	0	30 H
Ogden AP	41	5	6012	91	61	0	0	33 H
Price	39	5	....	91	60	0	0	33 H
Provo	40	6	5720	96	62	0	0	32 H
Richfield	38	5	....	91	60	0	0	34 H
St. George CO	37	21	....	101	65	0	0	33 H
Salt Lake City AP (S)	40	8	5990	95	62	0	0	32 H
Vernal AP	40	0	....	89	60	0	0	32 H
VERMONT								
Barre	44	-11	....	81	69	16	23	23 M
Burlington AP (S)	44	-7	8030	85	70	15	22	23 M
Rutland	43	-8	7440	84	70	16	23	23 M
VIRGINIA								
Charlottsville	38	18	4220	91	74	29	36	23 M
Danville AP	36	16	3510	92	73	21	28	21 M
Fredericksburg	38	14	....	93	75	31	38	21 M
Harrisonburg	38	16	....	91	72	17	24	23 M
Lynchburg AP	37	16	4166	90	74	30	37	21 M
Norfolk AP	36	22	3440	91	76	41	48	18 M
Petersburg	37	17	....	92	76	39	46	20 M
Richmond AP	37	17	3910	92	76	39	46	21 M
Roanoke AP	37	16	4150	91	72	17	24	23 M
Staunton	38	16	4307	91	72	17	24	23 M
Winchester	39	10	4780	90	74	30	37	21 M
WASHINGTON								
Aberdeen	47	28	5316	77	62	0	0	16 M
Bellingham AP	48	15	5420	77	65	2	9	19 M
Bremerton	47	25	5432	78	64	0	3	20 M
Ellensburg AP	47	6	....	91	64	0	0	34 H
Everett-Paine AFB	47	25	5940	76	64	0	7	20 M
Kennewick	46	11	....	96	67	0	0	30 H
Longview	46	24	5064	85	67	0	7	30 H
Moses Lake, Larson AFB	47	7	....	94	65	0	0	32 H
Olympia AP	47	22	5236	83	65	0	0	32 H
Port Angeles	48	27	....	69	61	0	3	18 M
Seattle-Boeing Fld	47	26	....	81	66	1	8	24 M
Seattle CO (S)	47	27	4424	82	66	0	7	19 M
Seattle-Tacoma AP (S)	47	26	5190	80	64	0	0	22 M
Spokane AP (S)	47	2	6770	90	63	0	0	28 M
Tacoma-McChord AFB	47	24	5287	82	65	0	2	22 M
Walla Walla AP	46	7	4800	94	66	0	0	27 H
Wenatchee	47	11	....	96	66	0	0	32 H
Yakima AP	46	5	5950	93	65	0	0	36 H

AP - Airport
CO - City Office
S - Solar Data Available

Table 1 (CONTINUED)

| Location | Latitude Degrees | WINTER | | SUMMER | | | | |
		97½% Design db	Heating D.D. Below 65°F	2½% Design db	Coincident Design wb	Grains Difference 55% RH	Grains Difference 50% RH	Daily Range
WEST VIRGINIA								
Beckley	37	4	5615	81	69	16	23	22 M
Bluefield AP	37	4	5000	81	69	16	23	22 M
Charleston AP	38	11	4510	90	73	24	31	20 M
Clarksburg	39	10	4590	90	73	24	31	21 M
Elkins AP	38	6	5680	84	70	16	23	22 M
Huntington CO	38	10	4340	91	74	29	36	22 M
Martinsburg AP	39	10	5231	90	74	30	37	21 M
Morgantown AP	39	8	5100	87	73	29	36	22 H
Parkersburg CO	39	11	4780	90	74	30	37	21 M
Wheeling	40	5	5220	86	71	19	26	21 M
WISCONSIN								
Appleton	44	-9	....	86	72	25	32	23 M
Ashland	46	-16	....	82	68	10	17	23 M
Beloit	42	-3	....	90	75	36	43	24 M
Eau Claire AP	44	-11	7970	89	73	26	33	23 M
Fond du Lac	43	-8	....	86	72	25	32	23 M
Green Bay AP	44	-9	8100	85	72	26	33	23 M
LaCrosse AP	43	-9	7530	88	73	28	35	22 M
Madison AP (S)	43	-7	7720	88	73	28	35	22 M
Manitowoc	44	-7	....	86	72	25	32	21 M
Marinette	45	-11	....	84	71	32	29	20 M
Milwaukee AP	43	-4	7470	87	73	29	36	21 M
Racine	42	-2	....	88	73	28	35	21 M
Sheboygan	43	-6	....	86	73	31	38	20 M
Stevens Point	43	-11	....	89	73	26	33	23 M
Waukesha	43	-5	....	87	73	29	36	22 M
Wausau AP	44	-12	8490	88	72	22	29	23 M
WYOMING								
Casper AP	42	-5	7510	90	57	0	0	31 H
Cheyene AP	41	-1	7370	86	58	0	0	30 H
Cody AP	44	-13	....	86	60	0	0	32 H
Evanston	41	-3	....	84	55	0	0	32 H
Lander AP (S)	42	-11	7870	88	61	0	0	32 H
Laramie AP (S)	41	-6	7560	81	56	0	0	28 H
Newcastle	43	-12	....	87	63	0	0	30 H
Rawlins	41	-4	....	83	57	0	0	40 H
Rock Springs AP	41	-3	8430	84	55	0	0	32 H
Sheridan AP	44	-8	7740	91	62	0	0	32 H
Torrington	42	-8	....	91	62	0	0	30 H

AP - Airport
CO - City Office
S - Solar Data Available

Table 1 (CONTINUED)

Location	Latitude Degrees	WINTER		SUMMER				
		97½% Design db	Heating D.D. Below 65°F	2½% Design db	Coincident Design wb	Grains Difference 55% RH	Grains Difference 50% RH	Daily Range
ALBERTA								
Calgary AP	51	-23	9703	81	61	0	0	25 M
Edmonton AP	53	-25	10268	82	65	0	2	23 M
Grande Prairie AP	55	-33	11129	80	63	0	0	23 M
Jasper	52	-26	10112	80	62	0	0	28 H
Lethbridge AP (S)	49	-22	8644	87	63	0	0	28 H
McMurray AP	56	-38	12462	82	65	0	2	26 H
Medicine Hat AP	50	-24	8852	90	65	0	0	28 H
Red Deer AP	52	-26	10302	81	64	0	0	25 M
BRITISH COLUMBIA								
Dawson Creek	55	-33	10800	79	63	0	0	26 H
Fort Nelson AP (S)	58	-40	10874	81	63	0	0	23 M
Kamloops CO	50	-15	6799	91	65	0	0	29 H
Nanaimo (S)	49	20	5554	80	65	0	4	21 M
New Westminister	49	18	....	81	67	5	12	19 M
Penticton AP	49	4	6522	89	67	0	0	31 H
Prince George AP (S)	53	-28	9755	80	62	0	0	26 H
Prince Rupert CO	54	2	2029	63	57	0	0	12 L
Trail	49	0	6711	89	65	0	0	33 H
Vancouver AP (S)	49	19	5515	77	66	8	15	17 M
Victoria CO	48	23	5579	73	62	0	0	16 M
MANITOBA								
Brandon	49	-27	10828	86	70	14	21	25 M
Churchill AP (S)	58	-39	16728	77	64	0	3	18 M
Dauphin AP	51	-28	....	84	70	16	23	23 M
Flin Flon	54	-37	12414	81	66	2	9	19 M
Portage la Prairie AP	49	-24	10800	86	72	25	32	22 M
The Pas AP (S)	53	-33	12281	82	67	4	11	20 M
Winnipeg AP (S)	49	-27	10679	86	71	19	26	22 M
NEW BRUNSWICK								
Campbellton CO	48	-14	....	82	67	4	11	21 M
Chatham AP	47	-10	....	85	68	5	12	22 M
Edmundston CO	47	-16	9796	83	68	8	15	21 M
Fredericton AP (S)	45	-11	8671	85	69	9	11	23 M
Moncton AP (S)	46	-8	8711	82	69	14	21	23 M
Saint John AP	45	-8	8453	77	65	3	10	19 M
NEWFOUNDLAND								
Corner Brook	48	0	8978	73	63	0	5	17 M
Gander AP	48	-1	9254	79	65	0	7	19 M
Goose Bay AP (S)	53	-24	11887	81	64	0	0	19 M
St. John's AP (S)	47	7	8991	75	65	6	13	18 M
Stephenville AP	48	4	8717	74	64	3	10	14 L
NORTHWEST TERR.								
Fort Smith AP (S)	60	-45	....	81	64	0	0	24 M
Frobisher AP (S)	63	-41	17876	63	51	0	0	14 L
Inuvik (S)	68	-53	....	77	60	0	0	21 M
Resolute AP (S)	74	-47	22673	54	46	0	0	10 L
Yellowknife AP	62	-46	15634	77	61	0	0	16 M
NOVA SCOTIA								
Amherst	45	-6	8400	81	68	11	18	21 M
Halifax AP (S)	44	5	7361	76	65	3	10	16 M
Kentville (S)	45	1	7792	83	68	8	15	22 M
New Glasgow	45	-5	....	79	68	14	21	20 M
Sydney AP	46	3	8049	80	68	12	19	19 M
Truro CO	45	-5	8226	80	69	18	25	22 M
Yarmouth AP	43	9	7340	72	64	6	13	15 L
ONTARIO								
Belleville	44	-7	7709	84	72	28	35	20 M
Chatham	42	3	6503	87	73	29	36	19 M
Cornwall	45	-9	8200	87	72	23	30	21 M
Hamilton	43	1	6821	86	72	25	32	21 M
Kaupuskasing AP (S)	49	-28	11560	83	69	13	20	23 M
Kenora AP	49	-28	10796	82	69	14	21	19 M
Kingston	44	-7	7724	84	72	28	35	20 M
Kitchener	43	-2	7566	85	72	26	33	23 M
London AP	43	0	7349	85	73	32	39	21 M
North Bay AP	46	-18	9677	81	67	6	13	20 M
Oshawa	43	-3	7600	86	72	25	32	20 M

AP - Airport
CO - City Office
S - Solar Data Available

Table 1 (CONTINUED)

Location	Latitude Degrees	WINTER 97½% Design db	WINTER Heating D.D. Below 65°F	SUMMER 2½% Design db	SUMMER Coincident Design wb	SUMMER Grains Difference 55% RH	SUMMER Grains Difference 50% RH	SUMMER Daily Range
Ottawa AP (S)	45	-13	8693	87	71	18	25	21 M
Owen Sound	44	-2	....	82	70	20	27	21 M
Peterborough	44	-9	8309	85	71	20	27	21 M
St. Catharines	43	3	6537	85	72	26	33	20 M
Sarnia	42	3	7061	86	72	25	32	19 M
Sault Ste. Marie AP	46	-13	9500	82	69	14	21	22 M
Sudbury AP	46	-19	9600	83	67	3	10	22 M
Thunder Bay AP	48	-24	10405	83	68	8	15	24 M
Timmins AP	48	-29	11400	84	68	6	13	25 M
Toronto AP (S)	43	-1	6827	87	72	23	30	20 M
Windsor AP	42	4	6579	88	73	28	35	20 M
PRINCE EDWARD ISLAND								
Charlottetown AP (S)	46	-4	8486	78	68	16	23	16 M
Summerside AP	46	-4	8440	79	68	14	21	16 M
QUEBEC								
Bagotville AP	48	-23	....	83	68	8	15	21 M
Chicoutimi	48	-22	10104	83	68	8	15	20 M
Drummondville	45	-14	8700	85	71	20	27	21 M
Granby	45	-14	8400	85	71	20	27	21 M
Hull	45	-14	8700	87	71	18	25	21 M
Megantic AP	45	-16	....	83	70	19	26	20 M
Montreal AP (S)	45	-10	8213	85	72	26	33	17 M
Quebec AP	46	-14	8937	84	70	16	23	20 M
Rimouski	48	-12	9906	79	66	4	11	18 M
St. Jean	45	-11	8500	86	72	25	32	20 M
St. Jeirome	45	-13	9285	86	71	19	26	23 M
Sept. Iles AP (S)	50	-21	....	73	61	0	0	17 M
Shawinigan	46	-14	9380	84	70	16	23	21 M
Sherbrooke CO	45	-21	8490	84	71	22	29	20 M
Thetford Mines	46	-14	9815	84	70	16	23	21 M
Trois Rivieres	46	-13	9306	85	70	15	22	23 M
Val d'Or AP	48	-27	11169	83	68	8	15	22 M
Valleyfield	45	-10	8300	86	72	25	32	20 M
SASKATCHEWAN								
Estevan AP	49	-25	9950	89	68	0	5	26 H
Moose Jaw AP	50	-25	9894	89	67	0	0	27 H
North Battleford AP	52	-30	11082	85	66	0	0	23 M
Prince Albert AP	53	-35	11630	84	66	0	2	25 M
Regina AP	50	-29	10806	88	68	0	7	26 H
Saskatoon AP (S)	52	-31	10856	86	66	0	0	26 H
Swift Current AP (S)	50	-25	9849	90	66	0	0	25 M
Yorkton AP	51	-30	11362	84	68	6	13	23 M
YUKON TERRITORY								
Whitehorse AP (S)	60	-43	12475	77	58	0	0	22 M

AP - Airport
CO - City Office
S - Solar Data Available

Table 2
Heat Transfer Multipliers (Heating)

No. 1 Single Pane Window

Winter Temperature Difference — HTM (Btuh per sq. ft.)

	20	25	30	35	40	45	50	55	60	65	70	75	80	85	90	95	U
Clear Glass																	
A. Wood Frame	19.8	24.8	29.7	34.7	39.6	44.6	49.5	54.5	59.4	64.4	69.3	74.3	79.2	84.2	89.1	94.1	.990
B. T.I.M. Frame	20.9	26.1	31.4	36.6	41.8	47.0	52.3	57.5	62.7	67.9	73.2	78.4	83.6	88.8	94.1	99.3	1.045
C. Metal Frame	23.1	28.9	34.7	40.4	46.2	52.0	57.8	63.5	69.3	75.1	80.9	86.6	92.4	98.2	104.0	109.7	1.155
Low Emittance Glass, e = 0.60																	
D. Wood Frame	18.4	23.0	27.5	32.1	36.7	41.3	45.9	50.5	55.1	59.7	64.3	68.9	73.4	78.0	82.6	87.2	.918
E. T.I.M. Frame	19.4	24.2	29.1	33.9	38.8	43.6	48.5	53.3	58.1	63.0	67.8	72.7	77.5	82.4	87.2	92.1	.969
F. Metal Frame	21.4	26.8	32.1	37.5	42.8	48.2	53.6	58.9	64.3	69.6	75.0	80.3	85.7	91.0	96.4	101.7	1.071
Low Emittance Glass, e = 0.40																	
G. Wood Frame	16.4	20.5	24.6	28.7	32.8	36.9	41.0	45.0	49.1	53.2	57.3	61.4	65.5	69.6	73.7	77.8	.819
H. T.I.M. Frame	17.3	21.6	25.9	30.3	34.6	38.9	43.2	47.5	51.9	56.2	60.5	64.8	69.2	73.5	77.8	82.1	.865
I. Metal Frame	19.1	23.9	28.7	33.4	38.2	43.0	47.8	52.6	57.3	62.1	66.9	71.7	76.4	81.2	86.0	90.8	.956
Low Emittance Glass, e = 0.20																	
J. Wood Frame	14.2	17.8	21.3	24.9	28.4	32.0	35.6	39.1	42.7	46.2	49.8	53.3	56.9	60.4	64.0	67.5	.711
K. T.I.M. Frame	15.0	18.8	22.5	26.3	30.0	33.8	37.5	41.3	45.0	48.8	52.5	56.3	60.0	63.8	67.5	71.3	.751
L. Metal Frame	16.6	20.7	24.9	29.0	33.2	37.3	41.5	45.6	49.8	53.9	58.1	62.2	66.4	70.5	74.7	78.8	.830

No. 2 Single Pane Window & Storm

Winter Temperature Difference — HTM (Btuh per sq. ft.)

	20	25	30	35	40	45	50	55	60	65	70	75	80	85	90	95	U
Clear Glass																	
A. Wood Frame	9.5	11.9	14.3	16.6	19.0	21.4	23.8	26.1	28.5	30.9	33.3	35.6	38.0	40.4	42.8	45.1	.475
B. T.I.M. Frame	10.5	13.1	15.8	18.4	21.0	23.6	26.3	28.9	31.5	34.1	36.8	39.4	42.0	44.6	47.3	49.9	.525
C. Metal Frame	13.0	16.3	19.5	22.8	26.0	29.3	32.5	35.8	39.0	42.3	45.5	48.8	52.0	55.3	58.5	61.8	.650
Low Emittance Glass																	
D. Wood Frame	8.4	10.5	12.5	14.6	16.7	18.8	20.9	23.0	25.1	27.2	29.3	31.4	33.4	35.5	37.6	39.7	.418
E. T.I.M. Frame	9.2	11.6	13.9	16.2	18.5	20.8	23.1	25.4	27.7	30.0	32.3	34.7	37.0	39.3	41.6	43.9	.462
F. Metal Frame	11.4	14.3	17.2	20.0	22.9	25.7	28.6	31.5	34.3	37.2	40.0	42.9	45.8	48.6	51.5	54.3	.572

No. 3 Double Pane Window

Winter Temperature Difference — HTM (Btuh per sq. ft.)

	20	25	30	35	40	45	50	55	60	65	70	75	80	85	90	95	U
Clear Glass																	
A. Wood Frame	11.0	13.8	16.5	19.3	22.0	24.8	27.6	30.3	33.1	35.8	38.6	41.3	44.1	46.8	49.6	52.3	.551
B. T.I.M. Frame	12.2	15.2	18.3	21.3	24.4	27.4	30.5	33.5	36.5	39.6	42.6	45.7	48.7	51.8	54.8	57.9	.609
C. Metal Frame	14.5	18.1	21.8	25.4	29.0	32.6	36.3	39.9	43.5	47.1	50.8	54.4	58.0	61.6	65.3	68.9	.725
Low Emittance Glass																	
D. Wood Frame	7.2	9.0	10.8	12.6	14.4	16.2	18.1	19.9	21.7	23.5	25.3	27.1	28.9	30.7	32.5	34.3	.361
E. T.I.M. Frame	8.0	10.0	12.0	14.0	16.0	18.0	20.0	21.9	23.9	25.9	27.9	29.9	31.9	33.9	35.9	37.9	.399
F. Metal Frame	9.5	11.9	14.3	16.6	19.0	21.4	23.8	26.1	28.5	30.9	33.3	35.6	38.0	40.4	42.8	45.1	.475
Adjustable Blind Between Panes																	
G. Wood Frame	4.8	5.9	7.1	8.3	9.5	10.7	11.9	13.1	14.3	15.4	16.6	17.8	19.0	20.2	21.4	22.6	.238
H. T.I.M. Frame	5.3	6.6	7.9	9.2	10.5	11.8	13.1	14.4	15.8	17.1	18.4	19.7	21.0	22.3	23.6	24.9	.263

Footnotes for this table are found on page 71.

Table 2 (Continued)

No. 4 Double Pane Window & Storm

	20	25	30	35	40	45	50	55	60	65	70	75	80	85	90	95	U
	— Winter Temperature Difference / HTM (Btuh per sq. ft.)																
Clear Glass																	
A. Wood Frame	6.8	8.5	10.2	11.9	13.7	15.4	17.1	18.8	20.5	22.2	23.9	25.6	27.3	29.0	30.7	32.4	.341
B. T.I.M. Frame	7.7	9.6	11.6	13.5	15.4	17.3	19.3	21.2	23.1	25.0	27.0	28.9	30.8	32.7	34.7	36.6	.385
C. Metal Frame	9.8	12.3	14.7	17.2	19.6	22.1	24.5	27.0	29.4	31.9	34.3	36.8	39.2	41.7	44.1	46.6	.490
Low Emittance Glass																	
D. Wood Frame	5.3	6.6	7.9	9.2	10.5	11.8	13.2	14.5	15.8	17.1	18.4	19.7	21.1	22.4	23.7	25.0	.263
E. T.I.M. Frame	5.9	7.4	8.9	10.4	11.9	13.4	14.9	16.3	17.8	19.3	20.8	22.3	23.8	25.2	26.7	28.2	.297
F. Metal Frame	7.6	9.5	11.3	13.2	15.1	17.0	18.9	20.8	22.7	24.6	26.5	28.4	30.2	32.1	34.0	35.9	.378

No. 5 Triple Pane Window

	20	25	30	35	40	45	50	55	60	65	70	75	80	85	90	95	U
Clear Glass																	
A. Wood Frame	7.6	9.5	11.4	13.3	15.2	17.1	19.0	20.9	22.8	24.7	26.6	28.5	30.4	32.3	34.2	36.1	.380
B. T.I.M. Frame	8.8	11.0	13.2	15.4	17.6	19.7	21.9	24.1	26.3	28.5	30.7	32.9	35.1	37.3	39.5	41.7	.439
C. Metal Frame	10.9	13.7	16.4	19.1	21.8	24.6	27.3	30.0	32.8	35.5	38.2	41.0	43.7	46.4	49.1	51.9	.546
Low Emittance Glass																	
D. Wood Frame	6.2	7.8	9.4	10.9	12.5	14.0	15.6	17.2	18.7	20.3	21.8	23.4	25.0	26.5	28.1	29.6	.312
E. T.I.M. Frame	7.2	9.0	10.8	12.6	14.4	16.2	18.0	19.8	21.6	23.4	25.2	27.0	28.8	30.6	32.4	34.2	.360
F. Metal Frame	9.0	11.2	13.4	15.7	17.9	20.2	22.4	24.6	26.9	29.1	31.4	33.6	35.8	38.1	40.3	42.6	.448
Clear Glass & Storm																	
G. Wood Frame	5.3	6.6	7.9	9.2	10.5	11.8	13.2	14.5	15.8	17.1	18.4	19.7	21.1	22.4	23.7	25.0	.263
H. T.I.M. Frame	6.2	7.8	9.3	10.9	12.4	14.0	15.5	17.1	18.6	20.2	21.7	23.2	24.8	26.4	27.9	29.5	.311
I. Metal Frame	7.6	9.5	11.3	13.2	15.1	17.0	18.9	20.8	22.7	24.6	26.5	28.4	30.2	32.1	34.0	35.9	.378

No. 6 Jalousie Windows

	20	25	30	35	40	45	50	55	60	65	70	75	80	85	90	95	U
A. Metal Frame-Single Glass	22.0	27.5	33.0	38.5	44.0	49.5	55.0	60.5	66.0	71.5	77.0	82.5	88.0	93.5	99.0	104.5	1.100
B. Metal Frame-Single Glass & Storm	10.0	12.5	15.0	17.5	20.0	22.5	25.0	27.5	30.0	32.5	35.0	37.5	40.0	42.5	45.0	47.5	.500

No. 7 Skylights

	20	25	30	35	40	45	50	55	60	65	70	75	80	85	90	95	U
Single; Clear Glass																	
A. Wood Frame	22.1	27.7	33.2	38.7	44.3	49.8	55.4	60.9	66.4	72.0	77.5	83.0	88.6	94.1	99.6	105.2	1.107
B. T.I.M. Frame	23.4	29.2	35.1	40.9	46.7	52.6	58.4	64.3	70.1	76.0	81.8	87.6	93.5	99.3	105.2	111.0	1.169
C. Metal Frame	25.8	32.3	38.7	45.2	51.7	58.1	64.6	71.0	77.5	83.9	90.4	96.9	103.3	109.8	116.2	122.7	1.292
Single; Plastic Dome																	
D. Wood Frame	20.7	25.9	31.1	36.2	41.4	46.6	51.8	56.9	62.1	67.3	72.5	77.6	82.8	88.0	93.2	98.3	1.035
E. T.I.M. Frame	21.9	27.3	32.8	38.2	43.7	49.2	54.6	60.1	65.6	71.0	76.5	81.9	87.4	92.9	98.3	103.8	1.093
F. Metal Frame	24.2	30.2	36.2	42.3	48.3	54.3	60.4	66.4	72.5	78.5	84.5	90.6	96.6	102.6	107.8	114.7	1.208
Double; Plastic Dome or Clear Glass																	
G. Wood Frame	13.3	16.6	20.0	23.3	26.6	29.9	33.3	36.6	39.9	43.2	46.6	49.9	53.2	56.5	59.9	63.2	.665
H. T.I.M. Frame	14.7	18.4	22.1	25.7	29.4	33.1	36.8	40.4	44.1	47.8	51.5	55.1	58.8	62.5	66.2	69.8	.735
I. Metal Frame	17.5	21.9	26.3	30.6	35.0	39.4	43.8	48.1	52.5	56.9	61.3	65.6	70.0	74.4	78.8	83.1	.875
Double; Low Emittance Glass																	
J. Wood Frame	9.9	12.4	14.8	17.3	19.8	22.2	24.7	27.2	29.6	32.1	34.6	37.1	39.5	42.9	44.5	46.9	.494
K. T.I.M. Frame	10.9	13.7	16.4	19.1	21.8	24.6	27.3	30.0	32.8	35.5	38.2	41.0	43.7	46.4	49.1	51.9	.546
L. Metal Frame	13.0	16.3	19.5	22.8	26.0	29.3	32.5	35.8	39.0	42.3	45.5	48.8	52.0	55.3	58.5	61.8	.650

All sub-tables show column headings "Winter Temperature Difference" and "HTM (Btuh per sq. ft.)".

Footnotes for this table are found on page 71.

Table 2 (Continued)

No. 8 Sliding Glass Doors

Winter Temperature Difference — HTM (Btuh per sq. ft.)

	20	25	30	35	40	45	50	55	60	65	70	75	80	85	90	95	U
Single Pane, Clear Glass																	
A. Wood Frame	19.8	24.8	29.7	34.7	39.6	44.6	49.5	54.5	59.4	64.4	69.3	74.3	79.2	84.2	89.1	94.1	.990
B. T.I.M. Frame	20.9	26.1	31.4	36.6	41.8	47.0	52.3	57.5	62.7	67.9	73.2	78.4	83.6	88.8	94.1	99.3	1.045
C. Metal Frame	23.1	28.9	34.7	40.4	46.2	52.0	57.8	63.5	69.3	75.1	80.9	86.6	92.4	98.2	104.0	109.7	1.155
Single Pane, Low "e" Glass																	
D. Wood Frame	16.4	20.5	24.6	28.7	32.8	36.9	41.0	45.0	49.1	53.2	57.3	61.4	65.5	69.6	73.7	77.8	.819
E. T.I.M. Frame	17.3	21.6	25.9	30.3	34.6	38.9	43.2	47.5	51.9	56.2	60.5	64.8	69.2	73.5	77.8	82.1	.865
F. Metal Frame	19.1	23.9	28.7	33.4	38.2	43.0	47.8	52.6	57.3	62.1	66.9	71.7	76.4	81.2	86.0	90.8	.956
Single Pane & Storm, Clear Glass																	
G. Wood Frame	9.5	11.9	14.3	16.6	19.0	21.4	23.8	26.1	28.5	30.9	33.3	35.6	38.0	40.4	42.8	45.1	.475
H. T.I.M. Frame	10.5	13.1	15.8	18.4	21.0	23.6	26.3	28.9	31.5	34.1	36.8	39.4	42.0	44.6	47.3	49.9	.525
I. Metal Frame	13.0	16.3	19.5	22.8	26.0	29.3	32.5	35.8	39.0	42.3	45.5	48.8	52.0	55.3	58.5	61.8	.650
Single Pane & Storm, Low "e" Glass																	
J. Wood Frame	8.4	10.5	12.5	14.6	16.7	18.8	20.9	23.0	25.1	27.2	29.3	31.4	33.4	35.5	37.6	39.7	.418
K. T.I.M. Frame	9.2	11.6	13.9	16.2	18.5	20.8	23.1	25.4	27.7	30.0	32.3	34.7	37.0	39.3	41.6	43.9	.462
L. Metal Frame	11.4	14.3	17.2	20.0	22.9	25.7	28.6	31.5	34.3	37.2	40.0	42.9	45.8	48.6	51.5	54.3	.572
Double Pane, Clear Glass																	
M. Wood Frame	11.0	13.8	16.5	19.3	22.0	24.8	27.6	30.3	33.1	35.8	38.6	41.3	44.1	46.8	49.6	52.3	.551
N. T.I.M. Frame	12.2	15.2	18.3	21.3	24.4	27.4	30.5	33.5	36.5	39.6	42.6	45.7	48.7	51.8	54.8	57.9	.609
O. Metal Frame	14.5	18.1	21.8	25.4	29.0	32.6	36.3	39.9	43.5	47.1	50.8	54.4	58.0	61.6	65.3	68.9	.725
Double Pane, Low "e" Glass																	
P. Wood Frame	7.2	9.0	10.8	12.6	14.4	16.2	18.1	19.9	21.7	23.5	25.3	27.1	28.9	30.7	32.5	34.3	.361
Q. T.I.M. Frame	8.0	10.0	12.0	14.0	16.0	18.0	20.0	21.9	23.9	25.9	27.9	29.9	31.9	33.9	35.9	37.9	.399
R. Metal Frame	9.5	11.9	14.3	16.6	19.0	21.4	23.8	26.1	28.5	30.9	33.3	35.6	38.0	40.4	42.8	45.1	.475
Triple Pane or Double Pane & Storm																	
S. Wood Frame	6.8	8.5	10.2	11.9	13.7	15.4	17.1	18.8	20.5	22.2	23.9	25.6	27.3	29.0	30.7	32.4	.341
T. T.I.M. Frame	7.7	9.6	11.6	13.5	15.4	17.3	19.3	21.2	23.1	25.0	27.0	28.9	30.8	32.7	34.7	36.6	.385
U. Metal Frame	9.8	12.3	14.7	17.2	19.6	22.1	24.5	27.0	29.4	31.9	34.3	36.8	39.2	41.7	44.1	46.6	.490
Triple or Double & Storm, Low "e"																	
V. Wood Frame	5.3	6.6	7.9	9.2	10.5	11.8	13.2	14.5	15.8	17.1	18.4	19.7	21.1	22.4	23.7	25.0	.263
W. T.I.M. Frame	5.9	7.4	8.9	10.4	11.9	13.4	14.9	16.3	17.8	19.3	20.8	22.3	23.8	25.2	26.7	28.2	.297
X. Metal Frame	7.6	9.5	11.3	13.2	15.1	17.0	18.9	20.8	22.7	24.6	26.5	28.4	30.2	32.1	34.0	35.9	.378

No. 9 French Doors

Winter Temperature Difference — HTM (Btuh per sq. ft.)

	20	25	30	35	40	45	50	55	60	65	70	75	80	85	90	95	U
Single Pane, Clear Glass																	
A. Wood Frame	18.7	23.4	28.1	32.7	37.4	42.1	46.8	51.4	56.1	60.8	65.5	70.1	74.8	79.5	84.2	88.8	.935
B. T.I.M. Frame	19.8	24.8	29.7	34.7	39.6	44.6	49.5	54.5	59.4	64.4	69.3	74.3	79.2	84.2	89.1	94.1	.990
C. Metal Frame	24.2	30.3	36.3	42.4	48.4	54.5	60.5	66.6	72.6	78.7	84.7	90.8	96.8	102.9	108.9	115.0	1.210
Single Pane, Low "e" Glass																	
D. Wood Frame	15.5	19.3	23.2	27.1	30.9	34.8	38.7	42.5	46.4	50.3	54.1	58.0	61.9	65.7	69.6	73.5	.774
E. T.I.M. Frame	16.4	20.5	24.6	28.7	32.8	36.9	41.0	45.0	49.1	53.2	57.3	61.4	65.5	69.6	73.7	77.8	.819
F. Metal Frame	20.0	25.0	30.0	35.0	40.0	45.0	50.1	55.1	60.1	65.1	70.1	75.1	80.1	85.1	90.1	95.1	1.001
Double Pane, Clear Glass																	
G. Wood Frame	10.4	13.1	15.7	18.3	20.9	23.5	26.1	28.7	31.3	33.9	36.5	39.2	41.8	44.4	47.0	49.6	.522
H. T.I.M. Frame	11.0	13.8	16.5	19.3	22.0	24.8	27.6	30.3	33.1	35.8	38.6	41.3	44.1	46.8	49.6	52.3	.551
I. Metal Frame	15.1	18.9	22.6	26.4	30.2	33.9	37.7	41.5	45.2	49.0	52.8	56.6	60.3	64.1	67.9	71.6	.754
Double Pane, Low "e" Glass																	
J. Wood Frame	6.8	8.6	10.3	12.0	13.7	15.4	17.1	18.8	20.5	22.2	23.9	25.7	27.4	29.1	30.8	32.5	.342
K. T.I.M. Frame	7.2	9.0	10.8	12.6	14.4	16.2	18.1	19.9	21.7	23.5	25.3	27.1	28.9	30.7	32.5	34.3	.361
L. Metal Frame	9.9	12.4	14.8	17.3	19.8	22.2	24.7	27.2	29.6	32.1	34.6	37.1	39.5	42.0	44.5	46.9	.494
Triple Pane, Clear Glass																	
M. Wood Frame	7.4	9.3	11.1	13.0	14.8	16.7	18.5	20.4	22.2	24.1	25.9	27.8	29.6	31.5	33.3	35.2	.371
N. T.I.M. Frame	7.8	9.8	11.7	13.7	15.6	17.6	19.5	21.5	23.4	25.4	27.3	29.3	31.2	33.2	35.1	37.1	.390
O. Metal Frame	11.7	14.6	17.6	20.5	23.4	26.3	29.3	32.2	35.1	38.0	41.0	43.9	46.8	49.7	52.7	55.6	.585
Triple Pane, Low "e" Glass																	
P. Wood Frame	5.1	6.4	7.7	9.0	10.3	11.5	12.8	14.1	15.4	16.7	18.0	19.2	20.5	21.8	23.1	24.4	.257
Q. T.I.M. Frame	5.4	6.8	8.1	9.5	10.8	12.2	13.5	14.9	16.2	17.6	18.9	20.3	21.6	23.0	24.3	25.7	.270
R. Metal Frame	8.1	10.1	12.2	14.2	16.2	18.2	20.3	22.3	24.3	26.3	28.4	30.4	32.4	34.4	36.5	38.5	.405

Footnotes are found on page 71.

Table 2 (Continued)

No. 10 Wood Doors

	Winter Temperature Difference																
	20	25	30	35	40	45	50	55	60	65	70	75	80	85	90	95	U
	HTM (Btuh per sq. ft.)																
A. Hollow Core	11.2	14.0	16.8	19.6	22.4	25.2	28.0	30.8	33.6	36.4	39.2	42.0	44.8	47.6	50.4	53.2	.560
B. Hollow Core & Wood Storm	6.6	8.3	9.9	11.6	13.2	14.9	16.5	18.2	19.8	21.5	23.1	24.8	26.4	28.1	29.7	31.4	.330
C. Hollow Core & Metal Storm	7.2	9.0	10.8	12.6	14.4	16.2	18.0	19.8	21.6	23.4	25.2	27.0	28.8	30.6	32.4	34.2	.360
D. Solid Core	9.2	11.5	13.8	16.1	18.4	20.7	23.0	25.3	27.6	29.9	32.2	34.5	36.8	39.1	41.4	43.7	.460
E. Solid Core & Wood Storm	5.8	7.3	8.7	10.2	11.6	13.1	14.5	16.0	17.4	18.9	20.3	21.8	23.2	24.7	26.1	27.6	.290
F. Solid Core & Metal Storm	6.4	8.0	9.6	11.2	12.8	14.4	16.0	17.6	19.2	20.8	22.4	24.0	25.6	27.2	28.8	30.4	.320
G. Panel	13.4	16.8	20.1	23.5	26.8	30.2	33.5	36.9	40.2	43.6	46.9	50.3	53.6	57.0	60.3	63.7	.670
H. Panel & Wood Storm	7.2	9.0	10.8	12.6	14.4	16.2	18.0	19.8	21.6	23.4	25.2	27.0	28.8	30.6	32.4	34.2	.360
I. Panel & Metal Storm	8.2	10.3	12.3	14.4	16.4	18.5	20.5	22.6	24.6	26.7	28.7	30.8	32.8	34.9	36.9	39.0	.410

No. 11 Metal Doors

	Winter Temperature Difference																
	20	25	30	35	40	45	50	55	60	65	70	75	80	85	90	95	U
	HTM (Btuh per sq. ft.)																
A. Fiberglass Core	11.8	14.8	17.7	20.7	23.6	26.6	29.5	32.5	35.4	38.4	41.3	44.3	47.2	50.2	53.1	56.1	.590
B. Fiberglass Core & Storm	7.3	9.2	11.0	12.8	14.7	16.5	18.4	20.2	22.0	23.9	25.7	27.5	29.4	31.2	33.0	34.9	.367
C. Polystyrene Core	9.4	11.8	14.1	16.5	18.8	21.2	23.5	25.9	28.2	30.6	32.9	35.3	37.6	40.0	42.3	44.7	.470
D. Polystyrene Core & Storm	6.3	7.9	9.5	11.1	12.7	14.3	15.9	17.4	19.0	20.6	22.2	23.8	25.4	26.9	28.5	30.1	.317
E. Urethane Core	3.8	4.8	5.7	6.7	7.6	8.6	9.5	10.5	11.4	12.4	13.3	14.3	15.2	16.2	17.1	18.1	.190
F. Urethane Core & Storm	3.4	4.3	5.1	6.0	6.8	7.7	8.5	9.4	10.2	11.1	11.9	12.8	13.6	14.5	15.3	16.2	.170

No. 12 Wood Frame Exterior Walls with Sheathing and Siding or Brick, or Other Exterior Finish.

	Winter Temperature Difference																
Cavity Insul Sheathing	20	25	30	35	40	45	50	55	60	65	70	75	80	85	90	95	U
	HTM (Btuh per sq. ft.)																
A. None ½" Gypsum Brd (R-0.5)	5.4	6.8	8.1	9.5	10.8	12.2	13.6	14.9	16.3	17.6	19.0	20.3	21.7	23.0	24.4	25.7	.271
B. None ½" Asphalt Brd (R-1.3)	4.3	5.4	6.5	7.6	8.7	9.8	10.8	11.9	13.0	14.1	15.2	16.3	17.4	18.4	19.5	20.6	.217
C. R-11 ½" Gypsum (R-0.5)	1.8	2.3	2.7	3.1	3.6	4.0	4.5	4.9	5.4	5.8	6.3	6.7	7.2	7.6	8.1	8.5	.090
D. R-11 ½" Asphalt Brd (R-1.3) R-11 ½" Bead Brd (R-1.8) R-13 ½" Gypsum Brd (R-0.5)	1.6	2.0	2.4	2.8	3.2	3.6	4.0	4.4	4.8	5.2	5.6	6.0	6.4	6.8	7.2	7.6	.080
E. R-11 ½" Extr Poly Brd (R-2.5) R-11 ¾" Bead Brd (R-2.7) R-13 ½" Asphalt Brd (R-1.3) R-13 ¾" Bead Brd (R-1.8)	1.5	1.9	2.3	2.6	3.0	3.4	3.8	4.1	4.5	4.9	5.3	5.6	6.0	6.4	6.8	7.1	.075
F. R-11 1" Bead Brd (R-3.6) R-11 ¾" Extr Poly Brd (R-3.8) R-13 ½" Extr Poly Brd (R-2.5) R-13 ¾" Bead Brd (R-2.7)	1.4	1.8	2.1	2.4	2.8	3.2	3.5	3.8	4.2	4.6	4.9	5.3	5.6	5.9	6.3	6.6	.070
G. R-13 ¾" Extr Poly Brd (R-3.8) R-13 1" Bead Brd (R-3.6)	1.3	1.6	2.0	2.3	2.6	2.9	3.3	3.6	3.9	4.2	4.6	4.9	5.2	5.5	5.9	6.2	.065
H. R-11 1" Extr Poly Brd (R-5.0) R-13 1" Extr Poly Brd (R-5.0) R-19 ½" Gypsum Brd (R-0.5)	1.2	1.5	1.8	2.1	2.4	2.7	3.0	3.3	3.6	3.9	4.2	4.5	4.8	5.1	5.4	5.7	.060
I. R-19 ½" Asphalt Brd (R-1.3) R-19 ½" Bead Brd (R-1.8)	1.1	1.4	1.6	1.9	2.2	2.5	2.8	3.0	3.3	3.6	3.8	4.1	4.4	4.7	4.9	5.2	.055
J. R-11 R-8 Sheathing R-13 R-8 Sheathing R-19 ½" or ¾" Extr Poly Brd R-19 ¾" or 1" Bead Brd	1.0	1.3	1.5	1.7	2.0	2.2	2.5	2.7	3.0	3.2	3.5	3.7	4.0	4.2	4.5	4.7	.050
K. R-19 1" Extr Poly Brd (R-5.0)	.9	1.1	1.3	1.6	1.8	2.0	2.2	2.5	2.7	2.9	3.1	3.4	3.6	3.8	4.0	4.3	.045
L. R-19 R-8 Sheathing	.8	1.0	1.2	1.4	1.6	1.8	2.0	2.2	2.4	2.6	2.8	3.0	3.2	3.4	3.6	3.8	.040
M. R-27 Wall	.7	.9	1.1	1.3	1.5	1.7	1.9	2.0	2.2	2.4	2.6	2.8	3.0	3.1	3.3	3.5	.037
N. R-30 Wall	.7	.8	1.0	1.2	1.3	1.5	1.7	1.8	2.0	2.1	2.3	2.5	2.6	2.8	3.0	3.1	.033
O. R-33 Wall	.6	.8	.9	1.1	1.2	1.4	1.5	1.7	1.8	2.0	2.1	2.3	2.4	2.6	2.7	2.9	.030

No. 13 Frame or Masonry Partitions Between a Conditioned and an Unconditioned Space.

Use HTM from Construction No. 12 or 14.
Select HTM for Actual Temperature Difference
Expected Across the Partition

Footnotes are found on page 71.

Table 2 (Continued)

No. 14 Masonry Walls, Block or Brick. Finished or Unfinished - Above Grade

	20	25	30	35	40	45	50	55	60	65	70	75	80	85	90	95	U
							HTM (Btuh per sq. ft.)										
A. 8" or 12" Block, No Insul., Unfin.	10.2	12.8	15.3	17.8	20.4	22.9	25.5	28.0	30.6	33.1	35.7	38.2	40.8	43.3	45.9	48.4	.510
B. 8" or 12" Block + R-5	2.9	3.6	4.3	5.0	5.8	6.5	7.2	7.9	8.6	9.4	10.1	10.8	11.5	12.2	13.0	13.7	.144
C. 8" or 12" Block + R-11	1.5	1.9	2.3	2.7	3.1	3.5	3.8	4.2	4.6	5.0	5.4	5.8	6.2	6.5	6.9	7.3	.077
D. 8" or 12" Block + R-19	1.0	1.2	1.4	1.7	1.9	2.2	2.4	2.6	2.9	3.1	3.4	3.6	3.8	4.1	4.3	4.6	.048
E. 4" Brick + 8" Block, No. Insul.	8.0	10.0	12.0	14.0	16.0	18.0	20.0	22.0	24.0	26.0	28.0	30.0	32.0	34.0	36.0	38.0	.400
F. 4" Brick + 8" Block + R-5	2.7	3.3	4.0	4.7	5.3	6.0	6.6	7.3	8.0	8.6	9.3	10.0	10.6	11.3	12.0	12.6	.133
G. 4" Brick + 8" Block + R-11	1.5	1.9	2.2	2.6	3.0	3.3	3.7	4.1	4.4	4.8	5.2	5.5	5.9	6.3	6.7	7.0	.074
H. 4" Brick + 8" Block + R-19	.9	1.2	1.4	1.6	1.9	2.1	2.3	2.6	2.8	3.1	3.3	3.5	3.8	4.0	4.2	4.5	.047

No. 15 Masonry Walls, Block or Brick. Finished or Unfinished - Below Grade*

	20	25	30	35	40	45	50	55	60	65	70	75	80	85	90	95	U
							HTM (Btuh per sq. ft.)										
Walls Extend 2'-5' Below Grade																	
A. 8" or 12" Block + No Insul.	2.5	3.1	3.7	4.4	5.0	5.6	6.2	6.9	7.5	8.1	8.7	9.4	10.0	10.6	11.2	11.9	.125
B. 8" or 12" Block + R-5	1.5	1.8	2.2	2.6	3.0	3.3	3.7	4.1	4.4	4.8	5.2	5.5	5.9	6.3	6.7	7.0	.074
C. 8" or 12" Block + R-11	1.0	1.3	1.5	1.8	2.0	2.3	2.6	2.8	3.1	3.3	3.6	3.8	4.1	4.3	4.6	4.8	.051
D. 8" or 12" Block + R-19	.7	.9	1.0	1.2	1.4	1.5	1.7	1.9	2.0	2.2	2.4	2.6	2.7	2.9	3.1	3.2	.034
Walls Extend More Than 5' Below Grade																	
E. 8" or 12" Block + No Insul.	1.7	2.2	2.6	3.0	3.5	3.9	4.3	4.8	5.2	5.6	6.1	6.5	6.9	7.4	7.8	8.2	.087
F. 8" or 12" Block + R-5	1.2	1.5	1.8	2.1	2.3	2.6	2.9	3.2	3.5	3.8	4.1	4.4	4.7	5.0	5.3	5.6	.059
G. 8" or 12" Block + R-11	.9	1.1	1.3	1.5	1.7	2.0	2.2	2.4	2.6	2.8	3.0	3.3	3.5	3.7	3.9	4.1	.043
H. 8" or 12" Block + R-19	.6	.8	.9	1.1	1.2	1.4	1.5	1.7	1.8	2.0	2.1	2.3	2.4	2.6	2.8	2.9	.031

No. 16 Ceilings Under a Ventilated Attic Space or Unheated Room

	20	25	30	35	40	45	50	55	60	65	70	75	80	85	90	95	U
							HTM (Btuh per sq. ft.)										
A. No Insulation	12.0	15.0	18.0	21.0	24.0	27.0	29.9	32.9	35.9	38.9	41.9	44.9	47.9	50.9	53.9	56.9	.599
B. R-7 Insulation	2.4	3.0	3.6	4.2	4.8	5.4	6.0	6.6	7.2	7.8	8.4	9.0	9.6	10.2	10.8	11.4	.120
C. R-11 Insulation	1.8	2.2	2.6	3.1	3.5	4.0	4.4	4.8	5.3	5.7	6.2	6.6	7.0	7.5	7.9	8.4	.088
D. R-19 Insulation	1.1	1.3	1.6	1.9	2.1	2.4	2.6	2.9	3.2	3.4	3.7	4.0	4.2	4.5	4.8	5.0	.053
E. R-22 Insulation	1.0	1.2	1.4	1.7	1.9	2.2	2.4	2.6	2.9	3.1	3.4	3.6	3.8	4.1	4.3	4.6	.048
F. R-26 Insulation	.8	1.0	1.1	1.3	1.5	1.7	1.9	2.1	2.3	2.5	2.7	2.8	3.0	3.2	3.4	3.6	.038
G. R-30 Insulation	.7	.8	1.0	1.2	1.3	1.5	1.6	1.8	2.0	2.1	2.3	2.5	2.6	2.8	3.0	3.1	.033
H. R-38 Insulation	.5	.7	.8	.9	1.0	1.2	1.3	1.4	1.6	1.7	1.8	2.0	2.1	2.2	2.3	2.5	.026
I. R-44 Insulation	.5	.6	.7	.8	.9	1.0	1.1	1.3	1.4	1.5	1.6	1.7	1.8	2.0	2.1	2.2	.023
J. R-57 Insulation	.3	.4	.5	.6	.7	.8	.8	.9	1.0	1.1	1.2	1.3	1.4	1.4	1.5	1.6	.017
K. Wood Decking, No Insulation	5.7	7.2	8.6	10.0	11.5	12.9	14.3	15.7	17.2	18.6	20.0	21.5	22.9	24.3	25.8	27.2	.287

No. 17 Roof on Exposed Beams or Rafters

	20	25	30	35	40	45	50	55	60	65	70	75	80	85	90	95	U
							HTM (Btuh per sq. ft.)										
A. 1½" Wood Decking, No Insul.	6.3	7.9	9.4	11.0	12.6	14.2	15.8	17.3	18.9	20.5	22.0	23.6	25.2	26.8	28.3	29.9	.315
B. 1½" Wood Decking + R-4	2.9	3.6	4.3	5.0	5.8	6.5	7.2	7.9	8.6	9.4	10.1	10.8	11.5	12.2	13.0	13.7	.144
C. 1½" Wood Decking + R-5	2.4	3.1	3.7	4.3	4.9	5.5	6.1	6.7	7.3	7.9	8.5	9.1	9.8	10.4	11.0	11.6	.122
D. 1½" Wood Decking + R-6	2.2	2.7	3.3	3.8	4.4	4.9	5.4	6.0	6.5	7.1	7.6	8.2	8.7	9.3	9.8	10.4	.109
E. 1½" Wood Decking + R-8	1.8	2.2	2.7	3.1	3.6	4.0	4.4	4.9	5.3	5.8	6.2	6.7	7.1	7.6	8.0	8.5	.089
F. 2" Shredded Wood Planks	4.3	5.4	6.5	7.6	8.7	9.8	10.8	11.9	13.0	14.1	15.2	16.3	17.4	18.4	19.5	20.6	.217
G. 3" Shredded Wood Plank	3.2	4.0	4.8	5.6	6.4	7.2	7.9	8.7	9.5	10.3	11.1	11.9	12.7	13.5	14.3	15.1	.159
H. 1½" Fiber Board Insulation	3.5	4.4	5.3	6.1	7.0	7.9	8.8	9.6	10.5	11.4	12.3	13.1	14.0	14.9	15.8	16.6	.175
I. 2" Fiber Board Insulation	2.8	3.5	4.2	4.9	5.6	6.3	7.0	7.7	8.4	9.1	9.8	10.5	11.2	11.9	12.6	13.3	.140
J. 3" Fiber Board Insulation	2.0	2.5	3.0	3.5	4.0	4.5	4.9	5.4	5.9	6.4	6.9	7.4	7.9	8.4	8.9	9.4	.099
K. 1½" Wood Decking + R-13	1.2	1.5	1.8	2.1	2.4	2.7	3.0	3.3	3.6	3.9	4.2	4.5	4.8	5.1	5.4	5.7	.060
L. 1½" Wood Decking + R-19	.8	1.0	1.2	1.4	1.6	1.8	2.0	2.3	2.5	2.7	2.9	3.1	3.3	3.5	3.7	3.9	.041

Footnotes are found on page 71.

Table 2 (Continued)

No. 18 Roof-Ceiling Combination

	Winter Temperature Difference																
	20	25	30	35	40	45	50	55	60	65	70	75	80	85	90	95	U
	HTM (Btuh per sq. ft.)																
A. No Insulation	6.2	7.7	9.2	10.8	12.3	13.9	15.4	16.9	18.5	20.0	21.6	23.1	24.6	26.2	27.7	29.3	.308
B. R-11 Batts	1.4	1.8	2.2	2.5	2.9	3.2	3.6	4.0	4.3	4.7	5.0	5.4	5.8	6.1	6.5	6.8	.072
C. R-19 Batts (2" x 8" Rafters)	1.0	1.2	1.5	1.7	2.0	2.2	2.4	2.7	2.9	3.2	3.4	3.7	3.9	4.2	4.4	4.7	.049
D. R-22 Batts (2" x 8" Rafters)	.9	1.1	1.3	1.6	1.8	2.0	2.2	2.5	2.7	2.9	3.1	3.4	3.6	3.8	4.0	4.3	.045
E. R-26 Batts (2" x 8" Rafters)	.8	1.0	1.2	1.4	1.6	1.8	2.0	2.2	2.4	2.6	2.8	3.0	3.2	3.4	3.6	3.8	.040
F. R-30" Batts (2" x 10" Rafters)	.7	.9	1.0	1.2	1.4	1.6	1.8	1.9	2.1	2.3	2.4	2.6	2.8	3.0	3.2	3.3	.035

No. 19 Floors Over an Unheated Basement, Enclosed Crawl Space* or Crawl Space with Closable Vents.

	Winter Temperature Difference																
	20	25	30	35	40	45	50	55	60	65	70	75	80	85	90	95	U
	HTM (Btuh per sq. ft.)																
A. Hardwood Floor + No Insulation	3.1	3.9	4.7	5.5	6.2	7.0	7.8	8.6	9.4	10.1	10.9	11.7	12.5	13.3	14.0	14.8	.312
B. Hardwood Floor + R-11	.8	1.0	1.2	1.4	1.6	1.8	2.0	2.2	2.4	2.6	2.8	3.0	3.2	3.4	3.6	3.8	.080
C. Hardwood Floor + R-13	.8	1.0	1.1	1.3	1.5	1.7	1.9	2.1	2.3	2.5	2.7	2.8	3.0	3.2	3.4	3.6	.076
D. Hardwood Floor + R-19	.5	.7	.8	.9	1.0	1.2	1.3	1.4	1.6	1.7	1.8	2.0	2.1	2.2	2.3	2.5	.052
E. Hardwood Floor + R-30	.4	.5	.6	.7	.8	.9	1.0	1.0	1.1	1.2	1.3	1.4	1.5	1.6	1.7	1.8	.037
F. Carpeted Floor + No Insulation	2.2	2.7	3.3	3.8	4.4	4.9	5.4	6.0	6.5	7.1	7.6	8.2	8.7	9.3	9.8	10.4	.218
G. Carpeted Floor + R-11	.7	.9	1.1	1.3	1.4	1.6	1.8	2.0	2.2	2.3	2.5	2.7	2.9	3.1	3.2	3.4	.071
H. Carpeted Floor + R-13	.7	.9	1.0	1.2	1.4	1.5	1.7	1.9	2.0	2.2	2.4	2.5	2.7	2.9	3.1	3.2	.068
I. Carpeted Floor + R-19	.5	.6	.7	.8	1.0	1.1	1.2	1.3	1.4	1.6	1.7	1.8	1.9	2.0	2.2	2.3	.048
J. Carpeted Floor + R-30	.4	.4	.5	.6	.7	.8	.9	1.0	1.1	1.2	1.3	1.3	1.4	1.5	1.6	1.7	.035

No. 20 Floors Over an Open Crawl Space or Garage

	Winter Temperature Difference																
	20	25	30	35	40	45	50	55	60	65	70	75	80	85	90	95	U
	HTM (Btuh per sq. ft.)																
A. Hardwood Floor + No Insulation	6.2	7.8	9.4	10.9	12.5	14.0	15.6	17.2	18.7	20.3	21.8	23.4	25.0	26.5	28.1	29.6	.312
B. Hardwood Floor + R-11	1.6	2.0	2.4	2.8	3.2	3.6	4.0	4.4	4.8	5.2	5.6	6.0	6.4	6.8	7.2	7.6	.080
C. Hardwood Floor + R-13	1.5	1.9	2.3	2.7	3.0	3.4	3.8	4.2	4.6	4.9	5.3	5.7	6.1	6.5	6.8	7.2	.076
D. Hardwood Floor + R-19	1.0	1.3	1.6	1.8	2.1	2.3	2.6	2.9	3.1	3.4	3.6	3.9	4.2	4.4	4.7	4.9	.052
E. Hardwood Floor + R-30	.7	.9	1.1	1.3	1.5	1.7	1.8	2.0	2.2	2.4	2.6	2.8	3.0	3.1	3.3	3.5	.037
F. Carpeted Floor + No Insulation	4.4	5.5	6.5	7.6	8.7	9.8	10.9	12.0	13.1	14.2	15.3	16.3	17.4	18.5	19.6	20.7	.218
G. Carpeted Floor + R-11	1.4	1.8	2.1	2.5	2.8	3.2	3.6	3.9	4.3	4.6	5.0	5.3	5.7	6.0	6.4	6.7	.071
H. Carpeted Floor + R-13	1.4	1.7	2.0	2.4	2.7	3.1	3.4	3.7	4.1	4.4	4.8	5.1	5.4	5.8	6.1	6.5	.068
I. Carpeted Floor + R-19	1.0	1.2	1.4	1.7	1.9	2.2	2.4	2.6	2.9	3.1	3.4	3.6	3.8	4.1	4.3	4.6	.048
J. Carpeted Floor + R-30	.7	.9	1.0	1.2	1.4	1.6	1.8	1.9	2.1	2.3	2.4	2.6	2.8	3.0	3.2	3.3	.035

No. 21 Basement Floors

	Winter Temperature Difference																
	20	25	30	35	40	45	50	55	60	65	70	75	80	85	90	95	U
	HTM (Btuh per sq. ft.)																
A. 2 or More Feet Below Grade	.5	.6	.7	.8	1.0	1.1	1.2	1.3	1.4	1.5	1.7	1.8	1.9	2.0	2.1	2.3	.024

No. 22 Concrete Slab on Grade

	Winter Temperature Difference																
	20	25	30	35	40	45	50	55	60	65	70	75	80	85	90	95	U
	HTM (Btuh per running linear foot of perimeter)																
A. No Edge Insulation	16.2	20.3	24.3	28.3	32.4	36.4	40.5	44.5	48.6	52.7	56.7	60.8	64.8	68.8	72.9	76.9	.810
B. 1" Edge Insulation, R = 5.0	8.2	10.3	12.3	14.3	16.4	18.4	20.5	22.5	24.6	26.6	28.7	30.8	32.8	34.8	36.9	38.9	.410
C. 1½" Edge Insulation, R = 8.0	5.4	6.8	8.1	9.4	10.8	12.1	13.5	14.8	16.2	17.5	18.9	20.2	21.6	22.9	24.3	25.6	.270
D. 2" Edge Insulation, R = 11.0	4.2	5.3	6.3	7.4	8.4	9.4	10.5	11.5	12.6	13.6	14.7	15.8	16.8	17.8	18.9	20.0	.210

No. 23 Concrete Slab with Perimeter Warm Air Duct System

	Winter Temperature Difference																
	20	25	30	35	40	45	50	55	60	65	70	75	80	85	90	95	U
	HTM (Btuh per running linear foot of perimeter)																
A. No Edge Insulation	38.0	47.5	57.0	66.5	76.0	85.5	95.0	104.5	114.0	123.5	133.0	142.5	152.0	161.5	171.0	180.5	1.90
B. 1" Edge Insulation, R = 5.0	22.8	28.5	34.2	39.9	45.6	51.3	57.0	62.7	68.4	74.1	79.8	85.5	91.2	96.9	102.6	108.3	1.14
C. 1½" Edge Insulation, R = 8.0	20.0	25.0	30.0	35.0	40.0	45.0	50.0	55.0	60.0	65.0	70.0	75.0	80.0	85.0	90.0	95.0	1.00
D. 2" Edge Insulation, R = 11.0	18.6	23.3	27.9	32.5	37.2	41.8	46.5	51.1	55.8	60.4	65.1	69.7	74.4	79.0	83.7	88.3	.930

Footnotes are found on page 71.

Footnotes to Table 2

1. Table 2 does not include any allowance for infiltration. U values are calculated for 15 mph wind (outdoor) velocity.

2. The HTM values for construction numbers 1 through 5 can be reduced to 85 percent of the values listed in this table if the window is equipped with an internal shading device which can be adjusted to provide a tight closure over the window area and which is installed as a permanent fixture.

3. "Low Emittance" refers to glass that has a coating or a composition which decreases the effective "U" value of the glass. Clear glass has an "e" value of 0.84 and the "e" value of low emittance glass can range from 0.10 to 0.60. The net "U" value of low emittance glass decreases as the "e" value of the coating decreases. The U values in this table are for a glass that has a "Low E" coating on one side of one pane of glass and do not include an allowance for multiple coatings.

   Three choices for the "e" value are included in construction number 1 because the "U" value for single pane construction is fairly sensitive to changes in the "e" value of the low emittance coating.

   A default "e" value of 0.40 was applied to construction numbers 2, 3, 4, and 5 because the "U" values for multiple pane construction are less sensitive to changes in the "e" value of the low emittance coating.

   When coated glass is installed but the "e" and "U" values are unknown, refer to the window manufacturers performance data, the builder or architect for the required information.

   If the exact "U" value for any given window or glass door is known, the heating HTM value can be precisely calculated by multiplying the "U" value by the winter design temperature difference.

4. The "U" values and "HTM" values for windows and glass doors have been adjusted for framing in accordance with the adjustment factors listed in Table 13C, Chapter 27 of the 1985 ASHRAE Fundamentals. The adjustment factors used for calculating the Manual J HTM values correspond to average value of the adjustment factors which are listed in the ASHRAE tables. These adjustment factors are intended to apply to windows and glass doors that have 80% to 90% glass and 20% to 10% frame. The "U" and "HTM" values for french doors have been adjusted for frame constructions which are wider than the frames used on windows and glass doors.

5. "T.I.M." refers to thermally improved metal frames which are manufactured with a thermal break between the indoor and outdoor frames.

6. The HTM values for skylights are based on Table 13C, Chapter 27 of the 1985 ASHRAE Fundamentals (Parts B and C). Averages of the Table 13C adjustment factors were used to account for the effects of framing. An "e value" of 0.40 was assumed for skylights that are constructed with low emittance glass.

7. The HTM values for wood and metal doors are based on the information provided in Tables 5A and 5B (Chapter 23 of the 1985 ASHRAE Fundamentals). Table 5B is supplemented by information from other sources. Wood doors include an allowance for a small pane of glass. Metal doors do not include an allowance for a small pane of glass.

8. Storm sash is assumed to be outdoor type with a 1" air space.

9. Wall U values include wood framing equal to 20% of the opaque wall area.

10. Ceiling U values include wood framing equal to 10% of the opaque ceiling area.

11. Floor U values include wood framing equal to 15% of the opaque floor area.

12. For walls below grade (construction number 15), the U values include an additional R value for the heat flow path through the soil. For walls which are two to five feet below grade, an additional R value of 4.85 was added to the wall R value. For walls which are five to eight feet below grade, an additional R value of 7.84 was added to the wall R value.

13. Because ground surface temperatures are somewhat higher than the winter outdoor design temperatures, effective "U" values equal to 85 percent of the calculated "U" values were used to calculate the HTM values for construction number 15 (below grade walls) and 21 (basement floors).

14. The temperature difference that was used to calculate the HTM values for floors over an unheated basement or enclosed crawl space (construction number 19) was assumed to be equal to 50 percent of the winter design temperature difference. For other temperature differences, the HTM can be calculated by multiplying the U value by the expected temperature difference.

15. Note - Masonry wall which has a floor that is less than two feet below grade should be considered as No. 14 (above grade). Use 15 A,B,C,D, if floor is 2 - 5 feet below grade. Use 15 E,F,G,H, if floor is 5 - 8 feet below grade. Calculate the wall area from ground level to the floor level for 15A through 15H.

Table 3A
Glass Heat Transfers Multipliers (Cooling)
No External Shade Screen
Clear Glass

Design Temperature Difference	Single Pane						Double Pane Single Pane & Low e Coating						Triple Pane Double Pane & Low e Coating					
DIRECTION WINDOW FACES	10	15	20	25	30	35	10	15	20	25	30	35	10	15	20	25	30	35
NO INTERNAL SHADING																		
N	23	27	31	35	39	43	19	21	23	25	27	29	17	18	19	20	21	22
NE and NW	56	60	64	68	72	76	47	49	51	53	55	57	43	44	45	46	47	48
E and W	81	85	89	93	97	101	68	70	72	74	76	78	62	63	64	65	66	67
SE and SW	70	74	78	82	86	90	59	61	63	65	67	69	53	54	55	56	57	58
S	40	44	48	52	56	60	34	36	38	40	42	44	30	31	32	33	34	35
DRAPERIES OR VENETIAN BLINDS																		
N	14	18	22	26	30	34	12	14	16	18	20	22	10	11	12	13	14	15
NE and NW	33	37	41	45	49	53	29	31	33	35	37	39	25	26	27	28	29	30
E and W	48	52	56	60	64	68	42	44	46	48	50	52	37	38	39	40	41	42
SE and SW	41	45	49	53	57	61	37	39	41	43	45	47	32	33	34	35	36	37
S	24	28	32	36	40	44	21	23	25	27	29	31	18	19	20	21	22	23
ROLLER SHADES — HALF DRAWN																		
N	17	21	25	29	33	37	16	18	20	22	24	26	14	15	16	17	18	19
NE and NW	41	45	49	53	57	61	38	40	42	44	46	48	34	35	36	37	38	39
E and W	60	64	68	72	76	80	55	57	59	61	63	65	49	50	51	52	53	54
SE and SW	52	56	60	64	68	72	47	49	51	53	55	57	42	43	44	45	46	47
S	30	34	38	42	46	50	27	29	31	33	35	37	24	25	26	27	28	29
AWNING, PORCHES, OR OTHER EXTERNAL SHADING																		
ALL DIRECTIONS	23	27	31	35	39	43	19	21	23	25	27	29	17	18	19	20	21	22

Tinted (Heat Absorbing) Glass

Design Temperature Difference	Single Pane						Double Pane Single Pane & Low e Coating						Triple Pane Double Pane & Low e Coating					
DIRECTION WINDOW FACES	10	15	20	25	30	35	10	15	20	25	30	35	10	15	20	25	30	35
NO INTERNAL SHADING																		
N	16	20	24	28	32	36	12	14	16	18	20	22	9	10	11	12	13	14
NE and NW	39	43	47	51	55	59	29	31	33	35	37	39	22	23	24	25	26	27
E and W	57	61	65	69	73	77	42	44	46	48	50	52	32	33	34	35	36	37
SE and SW	49	53	57	61	65	69	36	38	40	42	44	46	28	29	30	31	32	33
S	28	32	36	40	44	48	21	23	25	27	29	31	16	17	18	19	20	21
DRAPERIES OR VENETIAN BLINDS																		
N	12	16	20	24	28	32	9	11	13	15	17	19	6	7	8	9	10	11
NE and NW	30	34	38	42	46	50	22	24	26	28	30	32	15	16	17	18	19	20
E and W	43	47	51	55	59	63	31	33	35	37	39	41	21	22	23	24	25	26
SE and SW	37	41	45	49	53	57	27	29	31	33	35	37	18	19	20	21	22	23
S	21	25	29	33	37	41	15	17	19	21	23	25	10	11	12	13	14	15
ROLLER SHADES — HALF DRAWN																		
N	14	18	22	26	30	34	10	12	14	16	18	20	7	8	9	10	11	12
NE and NW	34	38	42	46	50	54	25	27	29	31	33	35	18	19	20	21	22	23
E and W	49	53	57	61	65	69	36	38	40	42	44	46	26	27	28	29	30	31
SE and SW	42	46	50	54	58	62	31	33	35	37	39	41	22	23	24	25	26	27
S	24	28	32	36	40	44	18	20	22	24	26	28	13	14	15	16	17	18
AWNING, PORCHES, OR OTHER EXTERNAL SHADING																		
ALL DIRECTIONS	16	20	24	28	32	36	12	14	16	18	20	22	9	10	11	12	13	14

Reflective Coated Glass

Design Temperature Difference	Single Pane						Double Pane Single Pane & Low e Coating						Triple Pane Double Pane & Low e Coating					
DIRECTION WINDOW FACES	10	15	20	25	30	35	10	15	20	25	30	35	10	15	20	25	30	35
NO INTERNAL SHADING																		
N	14	18	22	26	30	34	10	12	14	16	18	20	6	7	8	9	10	11
NE and NW	34	38	42	46	50	54	24	26	28	30	32	34	15	16	17	18	19	20
E and W	49	53	57	61	65	69	34	36	38	40	42	44	21	22	23	24	25	26
SE and SW	43	47	51	55	59	63	29	31	33	35	37	39	18	19	20	21	22	23
S	24	28	32	36	40	44	17	19	21	23	25	27	10	11	12	13	14	15
DRAPERIES OR VENETIAN BLINDS																		
N	11	15	19	23	27	31	8	10	12	14	16	18	5	6	7	8	9	10
NE and NW	28	32	36	40	44	48	20	22	24	26	28	30	12	13	14	15	16	17
E and W	40	44	48	52	56	60	30	32	34	36	38	40	17	18	19	20	21	22
SE and SW	35	39	43	47	51	55	26	28	30	32	34	36	15	16	17	18	19	20
S	20	24	28	32	36	40	15	17	19	21	23	25	9	10	11	12	13	14
ROLLER SHADES — HALF DRAWN																		
N	12	16	20	24	28	32	9	11	13	15	17	19	5	6	7	8	9	10
NE and NW	30	34	38	42	46	50	21	23	25	27	29	31	13	14	15	16	17	18
E and W	44	48	52	56	60	64	31	33	35	37	39	41	18	19	20	21	22	23
SE and SW	38	42	46	50	54	58	27	29	31	33	35	37	16	17	18	19	20	21
S	22	26	30	34	38	42	15	17	19	21	23	25	9	10	11	12	13	14
AWNING, PORCHES, OR OTHER EXTERNAL SHADING																		
ALL DIRECTIONS	14	18	22	26	30	34	10	12	14	16	18	20	6	7	8	9	10	11

Table 3B
Glass Heat Transfer Multipliers (Cooling)
External Shade Screen, Shading Coefficient = .15
Clear Glass

Design Temperature Difference	Single Pane						Double Pane Single Pane & Low e Coating						Triple Pane Double Pane & Low e Coating					
DIRECTION WINDOW FACES	10	15	20	25	30	35	10	15	20	25	30	35	10	15	20	25	30	35
NO INTERNAL SHADING																		
N	23	27	31	35	39	43	19	21	23	25	27	29	17	18	19	20	21	22
NE and NW	28	32	36	40	44	48	23	25	27	29	31	33	21	22	23	24	25	26
E and W	32	36	40	44	48	52	26	28	30	32	34	36	24	25	26	27	28	29
SE and SW	30	34	38	42	46	50	25	27	29	31	33	35	22	23	24	25	26	27
S	26	30	34	38	42	46	21	23	25	27	29	31	19	20	21	22	23	24
DRAPERIES OR VENETIAN BLINDS																		
N	14	18	22	26	30	34	12	14	16	18	20	22	10	11	12	13	14	15
NE and NW	17	21	25	29	33	37	15	17	19	21	23	25	12	13	14	15	16	17
E and W	19	23	27	31	35	39	17	19	21	23	25	27	14	15	16	17	18	19
SE and SW	18	22	26	30	34	38	16	18	20	22	24	26	13	14	15	16	17	18
S	16	20	24	28	32	36	13	15	17	19	21	23	11	12	13	14	15	16
ROLLER SHADES — HALF DRAWN																		
N	17	17	17	17	17	17	16	18	20	22	24	26	14	15	16	17	18	19
NE and NW	21	25	29	33	37	41	19	21	23	25	27	29	17	18	19	20	21	22
E and W	23	27	31	35	39	43	22	24	26	28	30	32	19	20	21	22	23	24
SE and SW	22	26	30	34	38	42	21	23	25	27	29	31	18	19	20	21	22	23
S	19	23	27	31	35	39	18	20	22	24	26	28	16	17	18	19	20	21

Tinted (Heat Absorbing) Glass

Design Temperature Difference	Single Pane						Double Pane Single Pane & Low e Coating						Triple Pane Double Pane & Low e Coating					
DIRECTION WINDOW FACES	10	15	20	25	30	35	10	15	20	25	30	35	10	15	20	25	30	35
NO INTERNAL SHADING																		
N	16	20	24	28	32	36	12	14	16	18	20	22	9	10	11	12	13	14
NE and NW	19	23	27	31	35	39	15	17	19	21	23	25	11	12	13	14	15	16
E and W	22	26	30	34	38	42	17	19	21	23	25	27	12	13	14	15	16	17
SE and SW	21	25	29	33	37	41	16	18	20	22	24	26	12	13	14	15	16	17
S	18	22	26	30	34	38	13	15	17	19	21	23	10	11	12	13	14	15
DRAPERIES OR VENETIAN BLINDS																		
N	12	16	20	24	28	32	9	11	13	15	17	19	6	7	8	9	10	11
NE and NW	15	19	23	27	31	35	11	13	15	17	19	21	7	8	9	10	11	12
E and W	17	21	25	29	33	37	12	14	16	18	20	22	8	9	10	11	12	13
SE and SW	16	20	24	28	32	36	12	14	16	18	20	22	8	9	10	11	12	13
S	13	17	21	25	29	33	10	12	14	16	18	20	7	8	9	10	11	12
ROLLER SHADES — HALF DRAWN																		
N	14	18	22	26	30	34	10	12	14	6	18	20	7	8	9	10	11	12
NE and NW	17	21	25	29	33	37	12	14	16	18	20	22	9	10	11	12	13	14
E and W	19	23	27	31	35	39	14	16	18	20	22	24	10	11	12	13	14	15
SE and SW	18	22	26	30	34	38	13	15	17	19	21	23	9	10	11	12	13	14
S	16	20	24	28	32	36	11	13	15	17	19	21	8	9	10	11	12	13

Reflective Coated Glass

Design Temperature Difference	Single Pane						Double Pane Single Pane & Low e Coating						Triple Pane Double Pane & Low e Coating					
DIRECTION WINDOW FACES	10	15	20	25	30	35	10	15	20	25	30	35	10	15	20	25	30	35
NO INTERNAL SHADING																		
N	14	18	22	26	30	34	10	12	14	16	18	20	6	7	8	9	10	11
NE and NW	17	21	25	29	33	37	12	14	16	18	20	22	7	8	9	10	11	12
E and W	19	23	27	31	35	39	14	16	18	20	22	24	8	9	10	11	12	13
SE and SW	18	22	26	30	34	38	13	15	17	19	21	23	8	9	10	11	12	13
S	16	20	24	28	32	36	11	13	15	17	19	21	7	8	9	10	11	12
DRAPERIES OR VENETIAN BLINDS																		
N	11	15	19	23	27	31	8	10	12	14	16	18	5	6	7	8	9	10
NE and NW	14	18	22	26	30	34	10	12	14	16	18	20	6	7	8	9	10	11
E and W	15	19	23	27	31	35	11	13	15	17	19	21	7	8	9	10	11	12
SE and SW	15	19	23	27	31	35	11	13	15	17	19	21	7	8	9	10	11	12
S	12	16	20	24	28	32	9	11	13	15	17	19	6	7	8	9	10	11
ROLLER SHADES — HALF DRAWN																		
N	12	16	20	24	28	32	9	11	13	15	17	19	5	6	7	8	9	10
NE and NW	15	19	23	27	31	35	11	13	15	17	19	21	6	7	8	9	10	11
E and W	17	21	25	29	33	37	12	14	16	18	20	22	7	8	9	10	11	12
SE and SW	16	20	24	28	32	36	12	14	16	18	20	22	7	8	9	10	11	12
S	14	18	22	26	30	34	10	12	14	16	18	20	6	7	8	9	10	11

Table 3C
Glass Heat Transfer Multipliers (Cooling)
External Shade Screen, Shading Coefficient = .25
Clear Glass

Design Temperature Difference	Single Pane						Double Pane — Single Pane & Low e Coating						Triple Pane — Double Pane & Low e Coating					
DIRECTION WINDOW FACES	10	15	20	25	30	35	10	15	20	25	30	35	10	15	20	25	30	35
NO INTERNAL SHADING																		
N	23	27	31	35	39	43	19	21	23	25	27	29	17	18	19	20	21	22
NE and NW	31	35	39	43	47	51	26	28	30	32	34	36	24	25	26	27	28	29
E and W	38	42	46	50	54	58	31	33	35	37	39	41	28	29	30	31	32	33
SE and SW	35	39	43	47	51	55	29	31	33	35	37	39	26	27	28	29	30	31
S	27	31	35	39	43	47	23	25	27	29	31	33	20	21	22	23	24	25
DRAPERIES OR VENETIAN BLINDS																		
N	14	18	22	26	30	34	12	14	16	18	20	22	10	11	12	13	14	15
NE and NW	19	23	27	31	35	39	16	18	20	22	24	26	14	15	16	17	18	19
E and W	23	27	31	35	39	43	20	22	24	26	28	30	17	18	19	20	21	22
SE and SW	21	25	29	33	37	41	18	20	22	24	26	28	16	17	18	19	20	21
S	17	21	25	29	33	37	14	16	18	20	22	24	12	13	14	15	16	17
ROLLER SHADES — HALF DRAWN																		
N	17	17	17	17	17	17	16	18	20	22	24	26	14	15	16	17	18	19
NE and NW	23	27	31	35	39	43	22	24	26	28	30	32	19	20	21	22	23	24
E and W	28	32	36	40	44	48	26	28	30	32	34	36	23	24	25	26	27	28
SE and SW	26	30	34	38	42	46	24	26	28	30	32	34	21	22	23	24	25	26
S	20	24	28	32	36	40	19	21	23	25	27	29	17	18	19	20	21	22

Tinted (Heat Absorbing) Glass

Design Temperature Difference	Single Pane						Double Pane — Single Pane & Low e Coating						Triple Pane — Double Pane & Low e Coating					
DIRECTION WINDOW FACES	10	15	20	25	30	35	10	15	20	25	30	35	10	15	20	25	30	35
NO INTERNAL SHADING																		
N	16	20	24	28	32	36	12	14	16	18	20	22	9	10	11	12	13	14
NE and NW	22	26	30	34	38	42	16	18	20	22	24	26	12	13	14	15	16	17
E and W	26	30	34	38	42	46	20	22	24	26	28	30	15	16	17	18	19	20
SE and SW	24	28	32	36	40	44	18	20	22	24	26	28	14	15	16	17	18	19
S	19	23	27	31	35	39	14	16	18	20	22	24	11	12	13	14	15	16
DRAPERIES OR VENETIAN BLINDS																		
N	12	16	20	24	28	32	9	11	13	15	17	19	6	7	8	9	10	11
NE and NW	17	21	25	29	33	37	12	14	16	18	20	22	8	9	10	11	12	13
E and W	20	24	28	32	36	40	15	17	19	21	23	25	10	11	12	13	14	15
SE and SW	18	22	26	30	34	38	14	16	18	20	22	24	9	10	11	12	13	14
S	14	18	22	26	30	34	11	13	15	17	19	21	7	8	9	10	11	12
ROLLER SHADES — HALF DRAWN																		
N	14	18	22	26	30	34	10	12	14	6	18	20	7	8	9	10	11	12
NE and NW	19	23	27	31	35	39	14	16	18	20	22	24	10	11	12	13	14	15
E and W	23	27	31	35	39	43	17	19	21	23	25	27	12	13	14	15	16	17
SE and SW	21	25	29	33	37	41	15	17	19	21	23	25	11	12	13	14	15	16
S	17	21	25	29	33	37	12	14	16	18	20	22	9	10	11	12	13	14

Reflective Coated Glass

Design Temperature Difference	Single Pane						Double Pane — Single Pane & Low e Coating						Triple Pane — Double Pane & Low e Coating					
DIRECTION WINDOW FACES	10	15	20	25	30	35	10	15	20	25	30	35	10	15	20	25	30	35
NO INTERNAL SHADING																		
N	14	18	22	26	30	34	10	12	14	16	18	20	6	7	8	9	10	11
NE and NW	19	23	27	31	35	39	14	16	18	20	22	24	8	9	10	11	12	13
E and W	23	27	31	35	39	43	16	18	20	22	24	26	10	11	12	13	14	15
SE and SW	21	25	29	33	37	41	15	17	19	21	23	25	9	10	11	12	13	14
S	17	21	25	29	33	37	12	14	16	18	20	22	7	8	9	10	11	12
DRAPERIES OR VENETIAN BLINDS																		
N	11	15	19	23	27	31	8	10	12	14	16	18	5	6	7	8	9	10
NE and NW	15	19	23	27	31	35	11	13	15	17	19	21	7	8	9	10	11	12
E and W	18	22	26	30	34	38	14	16	18	20	22	24	8	9	10	11	12	13
SE and SW	17	21	25	29	33	37	13	15	17	19	21	23	8	9	10	11	12	13
S	13	17	21	25	29	33	10	12	14	16	18	20	6	7	8	9	10	11
ROLLER SHADES — HALF DRAWN																		
N	12	16	20	24	28	32	9	11	13	15	17	19	5	6	7	8	9	10
NE and NW	17	21	25	29	33	37	12	14	16	18	20	22	7	8	9	10	11	12
E and W	20	24	28	32	36	40	15	17	19	21	23	25	8	9	10	11	12	13
SE and SW	19	23	27	31	35	39	14	16	18	20	22	24	8	9	10	11	12	13
S	15	19	23	27	31	35	11	13	15	17	19	21	6	7	8	9	10	11

Table 3D
Glass Heat Transfer Multipliers (Cooling)
External Shade Screen, Shading Coefficient = .35
Clear Glass

Design Temperature Difference	Single Pane						Double Pane Single Pane & Low e Coating						Triple Pane Double Pane & Low e Coating					
DIRECTION WINDOW FACES	10	15	20	25	30	35	10	15	20	25	30	35	10	15	20	25	30	35
NO INTERNAL SHADING																		
N	23	27	31	35	39	43	19	21	23	25	27	29	17	18	19	20	21	22
NE and NW	35	39	43	47	51	55	29	31	33	35	37	39	26	27	28	29	30	31
E and W	43	47	51	55	59	63	36	38	40	42	44	46	33	34	35	36	37	38
SE and SW	39	43	47	51	55	59	33	35	37	39	41	43	30	31	32	33	34	35
S	29	33	37	41	45	49	24	26	28	30	32	34	22	23	24	25	26	27
DRAPERIES OR VENETIAN BLINDS																		
N	14	18	22	26	30	34	12	14	16	18	20	22	10	11	12	13	14	15
NE and NW	21	25	29	33	37	41	18	20	22	24	26	28	15	16	17	18	19	20
E and W	26	30	34	38	42	46	23	25	27	29	31	33	19	20	21	22	23	24
SE and SW	23	27	31	35	39	43	21	23	25	27	29	31	18	19	20	21	22	23
S	18	22	26	30	34	38	15	17	19	21	23	25	13	14	15	16	17	18
ROLLER SHADES — HALF DRAWN																		
N	17	17	17	17	17	17	16	18	20	22	24	26	14	15	16	17	18	19
NE and NW	25	29	33	37	41	45	24	26	28	30	32	34	21	22	23	24	25	26
E and W	32	36	40	44	48	52	30	32	34	36	38	40	26	27	28	29	30	31
SE and SW	29	33	37	41	45	49	27	29	31	33	35	37	24	25	26	27	28	29
S	22	26	30	34	38	42	20	22	24	26	28	30	18	19	20	21	22	23

Tinted (Heat Absorbing) Glass

Design Temperature Difference	Single Pane						Double Pane Single Pane & Low e Coating						Triple Pane Double Pane & Low e Coating					
DIRECTION WINDOW FACES	10	15	20	25	30	35	10	15	20	25	30	35	10	15	20	25	30	35
NO INTERNAL SHADING																		
N	16	20	24	28	32	36	12	14	16	18	20	22	9	10	11	12	13	14
NE and NW	24	28	32	36	40	44	18	20	22	24	26	28	14	15	16	17	18	19
E and W	30	34	38	42	46	50	23	25	27	29	31	33	17	18	19	20	21	22
SE and SW	28	32	36	40	44	48	20	22	24	26	28	30	16	17	18	19	20	21
S	20	24	28	32	36	40	15	17	19	21	23	25	11	12	13	14	15	16
DRAPERIES OR VENETIAN BLINDS																		
N	12	16	20	24	28	32	9	11	13	15	17	19	6	7	8	9	10	11
NE and NW	18	22	26	30	34	38	14	16	18	20	22	24	9	10	11	12	13	14
E and W	23	27	31	35	39	43	17	19	21	23	25	27	11	12	13	14	15	16
SE and SW	21	25	29	33	37	41	15	17	19	21	23	25	10	11	12	13	14	15
S	15	19	23	27	31	35	11	13	15	17	19	21	7	8	9	10	11	12
ROLLER SHADES — HALF DRAWN																		
N	14	18	22	26	30	34	10	12	14	6	18	20	7	8	9	10	11	12
NE and NW	21	25	29	33	37	41	15	17	19	21	23	25	11	12	13	14	15	16
E and W	26	30	34	38	42	46	19	21	23	25	27	29	14	15	16	17	18	19
SE and SW	24	28	32	36	40	44	17	19	21	23	25	27	12	13	14	15	16	17
S	18	22	26	30	34	38	13	15	17	19	21	23	9	10	11	12	13	14

Reflective Coated Glass

Design Temperature Difference	Single Pane						Double Pane Single Pane & Low e Coating						Triple Pane Double Pane & Low e Coating					
DIRECTION WINDOW FACES	10	15	20	25	30	35	10	15	20	25	30	35	10	15	20	25	30	35
NO INTERNAL SHADING																		
N	14	18	22	26	30	34	10	12	14	16	18	20	6	7	8	9	10	11
NE and NW	21	25	29	33	37	41	15	17	19	21	23	25	9	10	11	12	13	14
E and W	26	30	34	38	42	46	18	20	22	24	26	28	11	12	13	14	15	16
SE and SW	24	28	32	36	40	44	17	19	21	23	25	27	10	11	12	13	14	15
S	18	22	26	30	34	38	12	14	16	18	20	22	7	8	9	10	11	12
DRAPERIES OR VENETIAN BLINDS																		
N	11	15	19	23	27	31	8	10	12	14	16	18	5	6	7	8	9	10
NE and NW	17	21	25	29	33	37	12	14	16	18	20	22	7	8	9	10	11	12
E and W	21	25	29	33	37	41	16	18	20	22	24	26	9	10	11	12	13	14
SE and SW	19	23	27	31	35	39	14	16	18	20	22	24	9	10	11	12	13	14
S	14	18	22	26	30	34	10	12	14	16	18	20	6	7	8	9	10	11
ROLLER SHADES — HALF DRAWN																		
N	12	16	20	24	28	32	9	11	13	15	17	19	5	6	7	8	9	10
NE and NW	18	22	26	30	34	38	13	15	17	19	21	23	8	9	10	11	12	13
E and W	23	27	31	35	39	43	17	19	21	23	25	27	10	11	12	13	14	15
SE and SW	21	25	29	33	37	41	15	17	19	21	23	25	9	10	11	12	13	14
S	16	20	24	28	32	36	11	13	15	17	19	21	6	7	8	9	10	11

Table 3E
Glass Heat Transfer Multipliers (Cooling)
External Shade Screen, Shading Coefficient = .45
Or Adjustable Blinds Between Glazing, (Double Pane Only)
Clear Glass

Design Temperature Difference	Single Pane						Double Pane Single Pane & Low e Coating						Triple Pane Double Pane & Low e Coating					
DIRECTION WINDOW FACES	10	15	20	25	30	35	10	15	20	25	30	35	10	15	20	25	30	35
NO INTERNAL SHADING																		
N	23	27	31	35	39	43	19	21	23	25	27	29	17	18	19	20	21	22
NE and NW	38	42	46	50	54	58	32	34	36	38	40	42	29	30	31	32	33	34
E and W	49	53	57	61	65	69	41	43	45	47	49	51	37	38	39	40	41	42
SE and SW	44	48	52	56	60	64	37	39	41	43	45	47	33	34	35	36	37	38
S	31	35	39	43	47	51	26	28	30	32	34	36	23	24	25	26	27	28
DRAPERIES OR VENETIAN BLINDS																		
N	14	18	22	26	30	34	12	14	16	18	20	22	10	11	12	13	14	15
NE and NW	23	27	31	35	39	43	20	22	24	26	28	30	17	18	19	20	21	22
E and W	29	33	37	41	45	49	26	28	30	32	34	36	22	23	24	25	26	27
SE and SW	26	30	34	38	42	46	23	25	27	29	31	33	20	21	22	23	24	25
S	19	23	27	31	35	39	16	18	20	22	24	26	14	15	16	17	18	19
ROLLER SHADES — HALF DRAWN																		
N	17	17	17	17	17	17	16	18	20	22	24	26	14	15	16	17	18	19
NE and NW	28	32	36	40	44	48	26	28	30	32	34	36	23	24	25	26	27	28
E and W	36	40	44	48	52	56	34	36	38	40	42	44	30	31	32	33	34	35
SE and SW	33	37	41	45	49	53	30	32	34	36	38	40	27	28	29	30	31	32
S	23	27	31	35	39	43	21	23	25	27	29	31	19	20	21	22	23	24

Tinted (Heat Absorbing) Glass

Design Temperature Difference	Single Pane						Double Pane Single Pane & Low e Coating						Triple Pane Double Pane & Low e Coating					
DIRECTION WINDOW FACES	10	15	20	25	30	35	10	15	20	25	30	35	10	15	20	25	30	35
NO INTERNAL SHADING																		
N	16	20	24	28	32	36	12	14	16	18	20	22	9	10	11	12	13	14
NE and NW	26	30	34	38	42	46	20	22	24	26	28	30	15	16	17	18	19	20
E and W	34	38	42	46	50	54	26	28	30	32	34	36	19	20	21	22	23	24
SE and SW	31	35	39	43	47	51	23	25	27	29	31	33	18	19	20	21	22	23
S	21	25	29	33	37	41	16	18	20	22	24	26	12	13	14	15	16	17
DRAPERIES OR VENETIAN BLINDS																		
N	12	16	20	24	28	32	9	11	13	15	17	19	6	7	8	9	10	11
NE and NW	20	24	28	32	36	40	15	17	19	21	23	25	10	11	12	13	14	15
E and W	26	30	34	38	42	46	19	21	23	25	27	29	13	14	15	16	17	18
SE and SW	23	27	31	35	39	43	17	19	21	23	25	27	11	12	13	14	15	16
S	16	20	24	28	32	36	12	14	16	18	20	22	8	9	10	11	12	13
ROLLER SHADES — HALF DRAWN																		
N	14	18	22	26	30	34	10	12	14	6	18	20	7	8	9	10	11	12
NE and NW	23	27	31	35	39	43	17	19	21	23	25	27	12	13	14	15	16	17
E and W	30	34	38	42	46	50	22	24	26	28	30	32	16	17	18	19	20	21
SE and SW	27	31	35	39	43	47	19	21	23	25	27	29	14	15	16	17	18	19
S	19	23	27	31	35	39	14	16	18	20	22	24	10	11	12	13	14	15

Reflective Coated Glass

Design Temperature Difference	Single Pane						Double Pane Single Pane & Low e Coating						Triple Pane Double Pane & Low e Coating					
DIRECTION WINDOW FACES	10	15	20	25	30	35	10	15	20	25	30	35	10	15	20	25	30	35
NO INTERNAL SHADING																		
N	14	18	22	26	30	34	10	12	14	16	18	20	6	7	8	9	10	11
NE and NW	23	27	31	35	39	43	16	18	20	22	24	26	10	11	12	13	14	15
E and W	30	34	38	42	46	50	21	23	25	27	29	31	13	14	15	16	17	18
SE and SW	27	31	35	39	43	47	19	21	23	25	27	29	11	12	13	14	15	16
S	19	23	27	31	35	39	13	15	17	19	21	23	8	9	10	11	12	13
DRAPERIES OR VENETIAN BLINDS																		
N	11	15	19	23	27	31	8	10	12	14	16	18	5	6	7	8	9	10
NE and NW	19	23	27	31	35	39	13	15	17	19	21	23	8	9	10	11	12	13
E and W	24	28	32	36	40	44	18	20	22	24	26	28	10	11	12	13	14	15
SE and SW	22	26	30	34	38	42	16	18	20	22	24	26	10	11	12	13	14	15
S	15	19	23	27	31	35	11	13	15	17	19	21	7	8	9	10	11	12
ROLLER SHADES — HALF DRAWN																		
N	12	16	20	24	28	32	9	11	13	15	17	19	5	6	7	8	9	10
NE and NW	20	24	28	32	36	40	14	16	18	20	22	24	9	10	11	12	13	14
E and W	26	30	34	38	42	46	19	21	23	25	27	29	11	12	13	14	15	16
SE and SW	24	28	32	36	40	44	17	19	21	23	25	27	10	11	12	13	14	15
S	17	21	25	29	33	37	12	14	16	18	20	22	7	8	9	10	11	12

Table 3F
Glass Heat Transfers Multipliers (Cooling)
Skylights
Clear Glass or Plastic

Design Temperature Difference	Single Pane						Double Pane						Triple Pane					
	10	15	20	25	30	35	10	15	20	25	30	35						
DIRECTION OPENING FACES	INCLINATION = 60 DEGREES																	
N	100	105	111	116	122	127	86	89	91	94	97	100						
NE and NW	128	134	139	145	150	156	110	113	116	118	121	124						
E and W	150	156	161	167	172	177	128	131	134	137	139	142						
SE and SW	141	146	152	157	162	168	121	123	126	129	132	134						
S	115	120	126	131	136	142	99	102	104	107	110	113						
	INCLINATION = 45 DEGREES																	
N	129	135	141	146	152	158	112	115	117	120	123	126						
NE and NW	153	158	164	170	175	181	132	134	137	140	143	146						
E and W	170	176	182	187	193	199	146	149	152	155	158	161						
SE and SW	163	168	174	180	185	191	140	143	146	148	151	154						
S	141	147	153	158	164	170	122	125	128	131	134	136						
	INCLINATION = 30 DEGREES																	
N	150	156	161	166	172	177	130	133	135	138	141	144						
NE and NW	167	172	177	183	188	194	144	147	149	152	155	158						
E and W	179	185	190	195	201	206	154	157	160	163	165	168						
SE and SW	174	179	184	190	195	201	150	153	155	158	161	164						
S	159	164	169	175	180	186	137	140	143	146	148	151						
	INCLINATION = 0 DEGREES																	
HORIZONTAL	160	164	168	172	176	180	139	141	143	145	147	149						

Tinted (Heat Absorbing) Glass or Opaque Plastic Skylights

Design Temperature Difference	Single Pane						Double Pane						Triple Pane					
	10	15	20	25	30	35	10	15	20	25	30	35						
DIRECTION OPENING FACES	INCLINATION = 60 DEGREES																	
N	70	75	81	86	92	97	55	58	60	63	66	69						
NE and NW	90	95	101	106	112	117	70	72	75	78	81	83						
E and W	106	111	116	122	127	133	81	84	86	89	92	95						
SE and SW	99	104	109	115	120	126	76	78	81	84	87	89						
S	80	86	91	97	102	108	63	65	68	71	74	76						
	INCLINATION = 45 DEGREES																	
N	91	96	102	107	113	119	71	74	77	80	83	86						
NE and NW	107	112	118	124	129	135	83	86	89	92	95	98						
E and W	120	125	131	136	142	148	93	95	98	101	104	107						
SE and SW	114	120	125	131	136	142	88	91	94	97	100	103						
S	99	105	110	116	122	127	78	81	83	86	89	92						
	INCLINATION = 30 DEGREES																	
N	105	110	116	121	127	132	83	86	89	91	94	97						
NE and NW	117	122	127	133	138	144	92	94	97	100	103	105						
E and W	126	131	136	142	147	153	98	101	104	106	109	112						
SE and SW	122	127	132	138	143	149	95	98	101	103	106	109						
S	111	116	122	127	133	138	88	90	93	96	99	101						
	INCLINATION = 0 DEGREES																	
HORIZONTAL	112	116	120	124	128	132	89	91	93	95	97	99						

Reflective Coated Skylights

Design Temperature Difference	Single Pane						Double Pane						Triple Pane					
	10	15	20	25	30	35	10	15	20	25	30	35						
DIRECTION OPENING FACES	INCLINATION = 60 DEGREES																	
N	52	58	63	69	74	79	41	43	46	49	52	54						
NE and NW	70	75	80	86	91	97	53	56	58	61	64	66						
E and W	83	88	93	99	104	110	62	64	67	70	72	75						
SE and SW	77	83	88	94	99	105	57	60	63	65	68	71						
S	61	66	72	77	83	88	47	49	52	55	58	60						
	INCLINATION = 45 DEGREES																	
N	65	70	76	81	87	93	51	54	57	59	62	65						
NE and NW	77	82	88	93	99	105	60	62	65	68	71	74						
E and W	85	91	96	102	107	113	67	69	72	75	78	81						
SE and SW	82	87	93	98	104	110	64	66	69	72	75	78						
S	71	76	82	88	93	99	56	59	62	64	67	70						
	INCLINATION = 30 DEGREES																	
N	76	81	86	92	97	103	60	63	65	68	71	74						
NE and NW	85	90	95	101	106	112	66	69	71	74	77	80						
E and W	92	97	102	108	113	119	71	74	76	79	82	85						
SE and SW	89	94	99	105	110	116	69	72	74	77	80	83						
S	81	86	91	97	102	108	63	66	68	71	74	77						
	INCLINATION = 0 DEGREES																	
HORIZONTAL	80	84	88	92	96	100	64	66	68	70	72	74						

Notes to Tables 3A — 3F

1. Table 3A is based on the information provided in Table 37, Chapter 26 of the 1985 ASHRAE Fundamentals (Table 37 is supplemented by information from other sources and some of the values in Table 37 have been slightly adjusted (plus or minus two Btuh/sq. ft. or less) to facilitate computerizing the Manual J Calculation Procedure.) Table 3A includes the effects of transmission and solar gains but does not include an allowance for infiltration.

2. The transmission and the solar heat gains are reduced when a low emittance coating is applied to a pane of clear, tinted or reflective glass. For a given type of glass, a single pane window with a low emittance coating is approximately equivalent to double pane window with no coating and a double pane window with a low emittance coating is approximately equal to a triple pane window with no coating. However, the exact heat gain depends on the emittance value of the coating and on the optical and thermal properties of the glass which receives the coating. Since there are many types of glass and many types of low emittance coatings the performance that is associated with specific combinations can vary considerably. Refer to the window manufacturer for detailed performance information.

3. In Tables 3B, 3C, 3D and 3E the solar radiation component of the total (radiation and transmission) heat gain is reduced by a percentage which is equal to the corresponding external shading coefficient.

4. The external shading coefficient (SC) for a shade screen is defined as the ratio of the solar gain for a externally shaded glass to the solar gain of a reference glass.

$$SC = \frac{\text{Solar Gain (1 Pane, Clear, No Internal Shade)} + \text{Shade Screen}}{\text{Solar Gain (1 Pane, Clear, No Internal Shade)}}$$

(The SC value is determined by tests conducted by the manufacturer of the shading device in accordance with industry standards. Refer to the manufacturers product literature or to Table 41, Chapter 27 of the 1985 ASHRAE Fundamentals for more information.)

5. Use the Table 3E HTM values (double glass only) for windows that include a venetian blind or louver as an integral part (installed in between two panes of fixed glass) of the window.

6. The horizontal skylight information in Table 3F is based on information which is published in Table 37, Chapter 26 of the 1985 ASHRAE Fundamentals. HTM values for inclined skylights are calculated by dividing the skylight unit area into vertical and horizontal components and then multiplying these components by the corresponding HTM values for horizontal and verticle (for a given exposure) glass and then adding these HTM component values.

Table 4
Heat Transfer Multipliers (Cooling)

No. 1 through 9 - Windows and Glass Doors - Refer to Table No. 3 for Summer HTM Values.

No. 10 - Wood Doors	Summer Temperature Difference and Daily Temperature Range												
	10		15			20			25		30	35	U
	L	M	L	M	H	L	M	H	M	H	H	H	
	HTM (Btuh per sq. ft.)												
A. Hollow Core	9.9	7.6	12.7	10.4	7.6	15.5	13.2	10.4	16.0	13.2	16.0	18.8	.560
B. Hollow Core & Wood Storm	5.8	4.5	7.5	6.1	4.5	9.1	7.8	6.1	9.4	7.8	9.4	11.1	.330
C. Hollow Core & Metal Storm	6.3	4.9	8.1	6.7	4.9	9.9	8.5	6.7	10.3	8.5	10.3	12.1	.360
D. Solid Core	8.1	6.3	10.4	8.6	6.3	12.7	10.9	8.6	13.2	10.9	13.2	15.5	.460
E. Solid Core & Wood Storm	5.1	3.9	6.6	5.4	3.9	8.0	6.8	5.4	8.3	6.8	8.3	9.7	.290
F. Solid Core & Metal Storm	5.6	4.4	7.2	6.0	4.4	8.8	7.6	6.0	9.2	7.6	9.2	10.8	.320
G. Panel	11.8	9.1	15.1	12.5	9.1	18.5	15.8	12.5	19.2	15.8	19.2	22.5	.670
H. Panel & Wood Storm	6.3	4.9	8.1	6.7	4.9	9.9	8.5	6.7	10.3	8.5	10.3	12.1	.360
I. Panel & Metal Storm	7.2	5.6	9.3	7.6	5.6	11.3	9.7	7.6	11.7	9.7	11.7	13.8	.410

No. 11 - Metal Doors	10		15			20			25		30	35	U
	L	M	L	M	H	L	M	H	M	H	H	H	
	HTM (Btuh per sq. ft.)												
A. Fiberglass Core	10.4	8.0	13.3	11.0	8.0	16.3	13.9	11.0	16.9	13.9	16.9	19.8	.590
B. Fiberglass Core & Storm	6.5	5.0	8.3	6.8	5.0	10.1	8.7	6.8	10.5	8.7	10.5	12.3	.367
C. Polystyrene Core	8.3	6.4	10.6	8.7	6.4	13.0	11.1	8.7	13.4	11.1	13.4	15.8	.470
D. Polystyrene Core & Storm	5.6	4.3	7.2	5.9	4.3	8.7	7.5	5.9	9.1	7.5	9.1	10.7	.317
E. Urethane Core	3.3	2.6	4.3	3.5	2.6	5.2	4.5	3.5	5.4	4.5	5.4	6.4	.190
F. Urethane Core & Storm	3.0	2.3	3.8	3.2	2.3	4.7	4.0	3.2	4.9	4.0	4.9	5.7	.170

No. 12 Wood Frame Exterior Walls With Sheathing and Siding or Brick Veneer or Other Exterior Finish.	10		15			20			25		30	35	U
	L	M	L	M	H	L	M	H	M	H	H	H	
	HTM (Btuh per sq. ft.)												
A. None ½" Gypsum Board (R-0.5)	4.8	3.7	6.1	5.0	3.7	7.5	6.4	5.0	7.8	6.4	7.8	9.1	.271
B. None ½" Asphalt Board (R-1.3)	3.8	3.0	4.9	4.0	3.0	6.0	5.1	4.0	6.2	5.1	6.2	7.3	.217
C. R-11 ½" Gypsum Board (R-0.5)	1.6	1.2	2.0	1.7	1.2	2.5	2.1	1.7	2.6	2.1	2.6	3.0	.090
D. R-11 ½" Asphalt Board (R-1.3) R-11 ½" Bead Brd. (R-1.8) R-13 ½" Gypsum Brd. (R-0.5)	1.4	1.1	1.8	1.5	1.1	2.2	1.9	1.5	2.3	1.9	2.3	2.7	.080
E. R-11 ½" Extr Poly Brd. (R-2.5) R-11 ¾" Bead Brd. (R-2.7) R-13 ½" Asphalt Brd. (R-1.3) R-13 ½" Bead Brd. (R-1.8)	1.3	1.0	1.7	1.4	1.0	2.1	1.8	1.4	2.1	1.8	2.1	2.5	.075
F. R-11 1" Bead Brd. (R-3.6) R-11 ¾" Extr Poly Brd. (R-3.8) R-13 ½" Extr Poly Brd (R-2.5) R-13 ¾" Bead Brd. (R-2.7)	1.2	1.0	1.6	1.3	1.0	1.9	1.7	1.3	2.0	1.7	2.0	2.4	.070
G. R-13 ¾" Extr Poly Brd. (R-3.8) R-13 1" Bead Brd (R-3.6)	1.1	.9	1.5	1.2	.9	1.8	1.5	1.2	1.9	1.5	1.9	2.2	.065
H. R-11 1" Extr Brd. (R-5.0) R-13 1" Extr Poly Brd. (R-5.0) R-19 ½" Gypsum Brd. (R-0.5)	1.1	.8	1.4	1.1	.8	1.7	1.4	1.1	1.7	1.4	1.7	2.0	.060
I. R-19 ½" Asphalt Brd. (R-1.3) R-19 ½" Bead Brd. (R-1.8)	1.0	.7	1.2	1.0	.7	1.5	1.3	1.0	1.6	1.3	1.6	1.8	.055
J. R-11 R-8 Sheathing R-13 R-8 Sheathing R-19 ½" or ¾" Extr Poly R-19 ¾" or 1" Bead Brd.	.9	.7	1.1	.9	.7	1.4	1.2	.9	1.4	1.2	1.4	1.7	.050
K. R-19 1" Extr Poly Brd (R-5.0)	.8	.6	1.0	.8	.6	1.2	1.1	.8	1.3	1.1	1.3	1.5	.045
L. R-19 R-8 Sheathing	.7	.5	.9	.7	.5	1.1	.9	.7	1.1	.9	1.1	1.3	.040
M. R-27 Wall	.7	.5	.8	.7	.5	1.0	.9	.7	1.1	.9	1.1	1.2	.037
N. R-30 Wall	.6	.4	.7	.6	.4	.9	.8	.6	.9	.8	.9	1.1	.033
O. R-33 Wall	.5	.4	.7	.6	.4	.8	.7	.6	.9	.7	.9	1.0	.030

Footnotes to Table 4 are found on page 84.

Table 4 (Continued)

No. 13 - Partitions Between Conditioned and Unconditioned Space - Wood Frame Partitions

	Summer Temperature Difference and Daily Temperature Range											
	10		15			20			25		30	35	U
	L	M	L	M	H	L	M	H	M	H	H	H	
	HTM (Btuh per sq. ft.)												
A. None ½" Gypsum Board (R-0.5)	2.4	1.4	3.8	2.7	1.4	5.1	4.1	2.7	5.4	4.1	5.4	6.8	.271
B. None ½" Asphalt Board (R-1.3)	2.0	1.1	3.0	2.2	1.1	4.1	3.3	2.2	4.3	.3.3	4.3	5.4	.217
C. R-11 ½" Gypsum Board (R-0.5)	.8	.4	1.3	.9	.4	1.7	1.3	.9	1.8	1.3	1.8	2.2	.090
D. R-11 ½" Asphalt Board (R-1.3) R-11 ½" Bead Brd. (R-1.8) R-13 ½" Gypsum Brd. (R-0.5)	.7	.4	1.1	.8	.4	1.5	1.2	.8	1.6	1.2	1.6	2.0	.080
E. R-11 ½" Extr Poly Brd. (R-2.5) R-11 ¾" Bead Brd. (R-2.7) R-13 ½" Asphalt Brd. (R-1.3) R-13 ½" Bead Brd. (R-1.8)	.7	.4	1.0	.8	.4	1.4	1.1	.8	1.5	1.1	1.5	1.9	.075
F. R-11 1" Bead Brd. (R-3.6) R-11 ¾" Extr Poly Brd. (R-3.8) R-13 ½" Extr Poly Brd. (R-2.5) R-13 ¾" Bead Brd. (R-2.7)	.6	.4	1.0	.7	.4	1.3	1.0	.7	1.4	1.0	1.4	1.8	.070
G. R-13 ¾" Extr Poly Brd. (R-3.8) R-13 1" Bead Brd (R-3.6)	.6	.3	.9	.6	.3	1.2	1.0	.6	1.3	1.0	1.3	1.6	.065
H. R-11 1" Extr Brd. (R-5.0) R-13 1" Extr Poly Brd. (R-5.0) R-19 ½" Gypsum Brd. (R-0.5)	.5	.3	.8	.6	.3	1.1	.9	.6	1.2	.9	1.2	1.5	.060
I. R-19 ½" Asphalt Brd. (R-1.3) R-19 ½" Bead Brd. (R-1.8)	.5	.3	.8	.5	.3	1.0	.8	.5	1.1	.8	1.1	1.4	.055
J. R-11 R-8 Sheathing R-13 R-8 Sheathing R-19 ½" or ¾" Extr Poly R-19 ¾" or 1" Bead Brd.	.4	.2	.7	.5	.2	.9	.7	.5	1.0	.7	1.0	1.2	.050
K. R-19 1" Extr Poly Brd. (R-5.0)	.4	.2	.6	.4	.2	.9	.7	.4	.9	.7	.9	1.1	.045
L. R-19 R-8 Sheathing	.4	.2	.6	.4	.2	.8	.6	.4	.8	.6	.8	1.0	.040

No. 13 - Partitions Between Conditioned & Unconditioned Space. Brick or Brick Partitions

	10		15			20			25		30	35	U
	L	M	L	M	H	L	M	H	M	H	H	H	
	HTM (Btuh per sq. ft.)												
M. 8" Brick, No Insul., Unfinished	1.3	0	3.8	1.8	0	6.4	4.3	1.8	6.9	4.3	6.9	9.4	.510
N. 8" Brick R-5	.4	0	1.1	.5	0	1.8	1.2	.5	1.9	1.2	1.9	2.7	.144
O. 8" Brick R-11	.2	0	.6	.3	0	1.0	.7	.3	1.0	.7	1.0	1.4	.077
P. 8" Brick R-19	.1	0	.4	.2	0	.6	.4	.2	.6	.4	.6	.9	.048
Q. 4" Brick 8" Block, No Insul.	1.0	0	3.0	1.4	0	5.0	3.4	1.4	5.4	3.4	5.4	7.4	.400
R. 4" Brick 8" Block R-5	.3	0	1.0	.5	0	1.7	1.1	.5	1.8	1.1	1.8	2.5	.133
S. 4" Brick 8" Block R-11	.2	0	.6	.3	0	.9	.6	.3	1.0	.6	1.0	1.4	.074
T. 4" Brick 8" Block R-19	.1	0	.4	.2	0	.6	.4	.2	.6	.4	.6	.9	.047

No. 14 - Masonry Walls, Block or Brick Finished or Unfinished - Above Grade

	10		15			20			25		30	35	U
	L	M	L	M	H	L	M	H	M	H	H	H	
	HTM (Btuh per sq. ft.)												
A. 8" or 12" Block, No Insul., Unfinished	5.3	3.2	7.8	5.8	3.2	10.4	8.3	5.8	10.9	8.3	10.9	13.4	.510
B. 8" or 12" Block + R-5	1.5	.9	2.2	1.6	.9	2.9	2.3	1.6	3.1	2.3	3.1	3.8	.144
C. 8" or 12" Block + R-11	.8	.5	1.2	.9	.5	1.6	1.3	.9	1.6	1.3	1.6	2.0	.077
D. 8" or 12" Block + R-19	.5	.3	.7	.5	.3	1.0	.8	.5	1.0	.8	1.0	1.3	.048
E. 4" Brick + 8" Block, No Insul.	4.1	2.5	6.1	4.5	2.5	8.1	6.5	4.5	8.5	6.5	8.5	10.5	.400
F. 4" Brick + 8" Block + R-5	1.4	.8	2.0	1.5	.8	2.7	2.2	1.5	2.8	2.2	2.8	3.5	.133
G. 4" Brick + 8" Block + R-11	.8	.5	1.1	.8	.5	1.5	1.2	.8	1.6	1.2	1.6	1.9	.074
H. 4" Brick + 8" Block + R-19	.5	.3	.7	.5	.3	1.0	.8	.5	1.0	.8	1.0	1.2	.047

Footnotes to Table 4 are found on page 84.

Table 4 (Continued)

Table 4 (Continued)

No. 15 - Masonry Walls, Block or Brick Below Grade -	Summer Temperature Difference and Daily Temperature Range												
	10		15			20			25		30	35	U
	L	M	L	M	H	L	M	H	M	H	H	H	
	HTM (Btuh per sq. ft.)												
All	0.0	0.0	0.0	0.0	0.0	0.0	0.0	0.0	0.0	0.0	0.0	0.0	

No. 16 - Ceilings Under a Ventilated Attic Space. Light Colored Roof	10		15			20			25		30	35	U
	L	M	L	M	H	L	M	H	M	H	H	H	
	HTM (Btuh per sq. ft.)												
A. No Insulation	13.1	11.4	15.3	13.5	11.4	17.5	15.7	13.5	17.9	15.7	17.9	20.1	.437
B. R-7 Insulation	3.4	2.9	3.9	3.5	2.9	4.5	4.0	3.5	4.6	4.0	4.6	5.2	.112
C. R-11 Insulation	2.5	2.2	2.9	2.6	2.2	3.3	3.0	2.6	3.4	3.0	3.4	3.8	.083
D. R-19 Insulation	1.6	1.4	1.9	1.6	1.4	2.1	1.9	1.6	2.2	1.9	2.2	2.4	.053
E. R-22 Insulation	1.4	1.2	1.7	1.5	1.2	1.9	1.7	1.5	2.0	1.7	2.0	2.2	.048
F. R-26 Insulation	1.1	1.0	1.3	1.2	1.0	1.5	1.4	1.2	1.6	1.4	1.6	1.7	.038
G. R-30 Insulation	1.0	.9	1.2	1.0	.9	1.3	1.2	1.0	1.4	1.2	1.4	1.5	.033
H. R-38 Insulation	.8	.7	.9	.8	.7	1.0	.9	.8	1.1	.9	1.1	1.2	.026
I. R-44 Insulation	.7	.6	.8	.7	.6	.9	.8	.7	.9	.8	.9	1.1	.023
J. R-57 Insulation	.5	.4	.6	.5	.4	.7	.6	.5	.7	.6	.7	.8	.017
K. Wood Decking, No Insulation	8.6	7.5	10.0	8.9	7.4	11.4	10.3	8.9	11.8	10.3	11.8	13.2	.287

No. 16 - Ceilings Under a Ventilated Attic (Btuh per sq. ft.) Dark Colored Roof	10		15			20			25		30	35	U
	L	M	L	M	H	L	M	H	M	H	H	H	
	HTM (Btuh per sq. ft.)												
A. No Insulation	16.6	14.9	18.8	17.0	14.9	21.0	19.2	17.0	21.4	19.2	21.4	23.6	.437
B. R-7 Insulation	4.3	3.8	4.8	4.4	3.8	5.4	4.9	4.4	5.5	4.9	5.5	6.0	.112
C. R-11 Insulation	3.2	2.8	3.6	3.2	2.8	4.0	3.7	3.2	4.1	3.7	4.1	4.5	.083
D. R-19 Insulation	2.0	1.8	2.3	2.1	1.8	2.5	2.3	2.1	2.6	2.3	2.6	2.9	.053
E. R-22 Insulation	1.8	1.6	2.1	1.9	1.6	2.3	2.1	1.9	2.4	2.1	2.4	2.6	.048
F. R-26 Insulation	1.4	1.3	1.6	1.5	1.3	1.8	1.7	1.5	1.9	1.7	1.9	2.1	.038
G. R-30 Insulation	1.3	1.1	1.4	1.3	1.1	1.6	1.5	1.3	1.6	1.5	1.6	1.8	.033
H. R-38 Insulation	1.0	.9	1.1	1.0	.9	1.2	1.1	1.0	1.3	1.1	1.3	1.4	.026
I. R-44 Insulation	.9	.8	1.0	.9	.8	1.1	1.0	.9	1.1	1.0	1.1	1.2	.023
J. R-57 Insulation	.6	.6	.7	.7	.6	.8	.7	.7	.8	.7	.8	.9	017
K. Wood Decking, No Insulation	10.9	9.8	12.3	11.2	9.8	13.8	12.6	11.2	14.0	12.6	14.0	15.5	.287

No. 17 - Roof on Exposed Beams or Rafters Light Colored Roof	10		15			20			25		30	35	U
	L	M	L	M	H	L	M	H	M	H	H	H	
	HTM (Btuh pr sq. ft.)												
A. 1½" Wood Decking, No Insul	8.8	7.6	10.3	9.1	7.6	11.8	10.6	9.1	12.1	10.6	12.1	13.5	.294
B. 1½" Wood Decking R-4	4.2	3.6	4.9	4.3	3.6	5.6	5.0	4.3	5.7	5.0	5.7	6.4	.140
C. 1½" Wood Decking R-5	3.6	3.1	4.2	3.7	3.1	4.8	4.3	3.7	4.9	4.3	4.9	5.5	.119
D. 1½" Wood Decking R-6	3.2	2.8	3.7	3.3	2.8	4.2	3.8	3.3	4.3	3.8	4.3	4.9	.106
E. 1½" Wood Decking R-8	2.6	2.3	3.1	2.7	2.3	3.5	3.2	2.7	3.6	3.2	3.6	4.0	.088
F. 2" Shredded Wood Planks	6.2	5.4	7.2	6.4	5.4	8.3	7.5	6.4	8.5	7.5	8.5	9.5	.207
G. 3" Shredded Wood Planks	4.6	4.0	5.4	4.3	4.0	6.2	5.5	4.8	6.3	5.5	6.3	7.1	.154
H. 1½" Fiber Board Insulation	5.1	4.4	5.9	5.2	4.4	6.8	6.1	5.2	6.9	6.1	6.9	7.8	.169
I. 2" Fiber Board Insulation	4.0	3.5	4.7	4.2	3.5	5.4	4.9	4.2	5.5	4.9	5.5	6.2	.135
J. 3" Fiber Board Insulation	2.9	2.5	3.4	3.0	2.5	3.9	3.5	3.0	4.0	3.5	4.0	4.5	.097
K. 1½" Wood Decking R-13	1.8	1.6	2.1	1.9	1.6	2.4	2.2	1.9	2.5	2.2	2.5	2.8	.060
L. 1½" Wood Decking R-19	1.2	1.1	1.4	1.3	1.1	1.6	1.5	1.3	1.7	1.5	1.7	1.9	.041

Footnotes to Table 4 are found on page 84.

Table 4 (Continued)

No. 17 - Roof on Exposed Beams or Rafters — Dark Colored Roof

	10		15			20			25		30	35	U
	L	M	L	M	H	L	M	H	M	H	H	H	
						HTM (Btuh per sq. ft.)							
A. 1½" Wood Decking, No Insul.	11.2	10.0	12.6	11.5	10.0	14.1	12.9	11.5	14.4	12.9	14.4	15.9	.294
B. 1½" Wood Decking R-4	5.3	4.8	6.0	5.5	4.8	6.7	6.2	5.5	6.9	6.2	6.9	7.6	.140
C. 1½" Wood Decking R-5	4.5	4.0	5.1	4.6	4.0	5.7	5.2	4.6	5.8	5.2	5.8	6.4	.119
D. 1½" Wood Decking R-6	4.0	3.6	4.6	4.1	3.6	5.1	4.7	4.1	5.2	4.7	5.2	5.7	.106
E. 1½" Wood Decking R-8	3.3	3.0	3.8	3.4	3.0	4.2	3.9	3.4	4.3	3.9	4.3	4.8	.088
F. 2" Shredded Wood Planks	7.9	7.0	8.9	8.1	7.0	9.9	9.1	8.1	10.1	9.1	10.1	11.2	.207
G. 3" Shredded Wood Planks	5.9	5.2	6.6	6.0	5.2	7.4	6.8	6.0	7.5	6.8	7.5	8.3	.154
H. 1½" Fiber Board Insulation	6.4	5.7	7.3	6.6	5.7	8.1	7.4	6.6	8.3	7.4	8.3	9.1	.169
I. 2" Fiber Board Insulation	5.1	4.6	5.8	5.3	4.6	6.5	5.9	5.3	6.6	5.9	6.6	7.3	.136
J. 3" Fiber Board Insulation	3.7	3.3	4.2	3.8	3.3	4.7	4.3	3.8	4.8	4.3	4.8	5.2	.097
K. 1½" Wood Decking R-13	2.3	2.0	2.6	2.3	2.0	2.9	2.6	2.3	2.9	2.6	2.9	3.2	.060
L. 1½" Wood Decking R-19	1.6	1.4	1.8	1.6	1.4	2.0	1.8	1.6	2.0	1.8	2.0	2.2	.041

No. 18 - Roof-Ceiling Combination - Light Colored Roof

	10		15			20			25		30	35	U
	L	M	L	M	H	L	M	H	M	H	H	H	
						HTM (Btuh per sq. ft.)							
A. No Insulation	8.6	7.5	10.0	8.9	7.5	11.5	10.3	8.9	11.8	10.3	11.8	13.2	.287
B. R-11 Batts	2.2	1.9	2.5	2.2	1.9	2.9	2.6	2.2	3.0	2.6	3.0	3.3	.072
C. R-19 Batts	1.5	1.3	1.7	1.5	1.3	2.0	1.8	1.5	2.0	1.8	2.0	2.3	.049
D. R-22 Batts (2" x 8" Rafters)	1.3	1.2	1.6	1.4	1.2	1.8	1.6	1.4	1.8	1.6	1.8	2.1	.045
E. R-26 Batts (2" x 8" Rafters)	1.2	1.0	1.4	1.2	1.0	1.6	1.4	1.2	1.6	1.4	1.6	1.8	.040
F. R-30 Batts (2" x 10" Rafters)	1.0	.9	1.2	1.1	.9	1.4	1.3	1.1	1.4	1.3	1.4	1.6	.035

No. 18 - Roof-Ceiling Combination Dark Colored Roof

	10		15			20			25		30	35	U
	L	M	L	M	H	L	M	H	M	H	H	H	
						HTM (Btuh per sq. ft.)							
A. No Insulation	10.9	9.8	12.3	11.2	9.8	13.8	12.6	11.2	14.1	12.6	14.1	15.5	.287
B. R-11 Batts	2.7	2.4	3.1	2.8	2.4	3.5	3.2	2.8	3.5	3.2	3.5	3.9	.072
C. R-19 Batts	1.9	1.7	2.1	1.9	1.7	2.4	2.2	1.9	2.4	2.2	2.4	2.6	.049
D. R-22 Batts (2" x 8" Rafters)	1.7	1.5	1.9	1.8	1.5	2.2	2.0	1.8	2.2	2.0	2.2	2.4	.045
E. R-26 Batts (2" x 8" Rafters)	1.5	1.4	1.7	1.6	1.4	1.9	1.8	1.6	2.0	1.8	2.0	2.2	.040
F. R-30 Batts (2" x 10" Rafters)	1.3	1.2	1.5	1.4	1.2	1.7	1.5	1.4	1.7	1.5	1.7	1.9	.035

No. 19 - Floors Over a Basement or Enclosed Crawl Space

	10		15			20			25		30	35	U
	L	M	L	M	H	L	M	H	M	H	H	H	
						HTM (Btuh per sq. ft.)							
All	0.0	0.0	0.0	0.0	0.0	0.0	0.0	0.0	0.0	0.0	0.0	0.0	

No. 20 - Floors Over an Open Crawl Space or Garage

	10		15			20			25		30	35	U
	L	M	L	M	H	L	M	H	M	H	H	H	
						HTM (Btuh per sq. ft.)							
A. Hardwood Floor, No Insulation	3.5	1.9	5.4	3.9	1.9	7.3	5.8	3.9	7.7	5.8	7.7	9.6	.386
B. Hardwood Floor R-11	.8	.4	1.2	.8	.4	1.6	1.3	.8	1.7	1.3	1.7	2.1	.084
C. Hardwood Floor R-13	.7	.4	1.1	.8	.4	1.4	1.1	.8	1.5	1.1	1.5	1.9	.076
D. Hardwood Floor R-19	.5	.3	.8	.5	.3	1.0	.8	.5	1.1	.8	1.1	1.3	.054
E. Hardwood Floor R-30	.3	.2	.5	.4	.2	.7	.6	.4	.7	.6	.7	.9	.037
F. Carpet Floor No Insulation	2.3	1.3	3.5	2.5	1.3	4.8	3.8	2.5	5.1	3.8	5.1	6.3	.253
G. Carpet Floor R-11	.7	.4	1.0	.8	.4	1.4	1.1	.8	1.5	1.1	1.5	1.9	.075
H. Carpet Floor R-13	.6	.3	1.0	.7	.3	1.3	1.0	.7	1.4	1.0	1.4	1.7	.068
I. Carpet Floor R-19	.4	.2	.7	.5	.2	.9	.7	.5	1.0	.7	1.0	1.2	.050
J. Carpet Floor R-30	.3	.2	.5	.4	.2	.7	.5	.4	.7	.5	.7	.9	.035

No. 21 - 23 Basement Floors, Concrete Slab on Grade

	10		15			20			25		30	35	U
	L	M	L	M	H	L	M	H	M	H	H	H	
						HTM							
All	0.0	0.0	0.0	0.0	0.0	0.0	0.0	0.0	0.0	0.0	0.0	0.0	

Notes to Table 4 — Heat Transfer Multipliers (Cooling)

1. The HTM shown in this table do not include credit for infiltration. Refer to Table 5 for summer infiltration calculation procedure.
2. Wall U values include wood framing equal to 20% of the opaque wall area.
3. Ceiling U values include wood framing equal to 10% of the opaque ceiling area.
4. Floor U values include wood framing equal to 15% of the opaque floor area.
5. Summer HTM values include the effects of solar radiation and thermal mass!

Table 5

Infiltration Evaluation

Winter Air Changes Per Hour

Floor Area	900 or less	900-1500	1500-2100	over 2100
Best	0.4	0.4	0.3	0.3
Average	1.2	1.0	0.8	0.7
Poor	2.2	1.6	1.2	1.0

For each fire place add:		Best	Average	Poor
		0.1	0.2	0.6

Summer Air Changes Per Hour

Floor Area	900 or less	900-1500	1500-2100	over 2100
Best	0.2	0.2	0.2	0.2
Average	0.5	0.5	0.4	0.4
Poor	0.8	0.7	0.6	0.5

Envelope Evaluation

Best - Continuous infiltration barrier, all cracks and penetrations sealed, tested leakage of windows and doors less then 0.25 CFM per running foot of crack, vents and exhaust fans dampered, recessed ceiling lights gasketed or taped, no combustion air required or combustion air from outdoors, no duct leakage.

Average - Plastic vapor barrier, major cracks and penetrations sealed, tested leakage of windows and doors between 0.25 and 0.50 CFM per running foot of crack, electrical fixtures which penetrate the envelope not taped or gasketed, vents and exhaust fans dampered, combustion air from indoors, intermittent ignition and flue damper, some duct leakage to unconditioned space.

Poor - No infiltration barrier or plastic vapor barrier, no attempt to seal cracks and penetrations, tested leakage of windows and doors greater than 0.50 CFM per running foot of crack, vents and exhaust fans not dampered, combustion air from indoors, standing pilot, no flue damper, considerable duct leakage to unconditioned space.

Fireplace Evaluation

Best - Combustion air from outdoors, tight glass doors and damper.
Average - Combustion air from indoors, tight glass doors or damper.
Poor - Combustion air from indoors, no glass doors or damper.

Notes To Table 5

1. One, two or three story, or split level; any wind exposure.
2. Floor plan aspect ratio between 1:1 and 3:1.
3. Glass plus door areas between 10% and 30% of the wall area.
4. Allowance for one kitchen and two bathroom exhaust fans, dryer vent, recessed lighting fixtures, pipe and duct penetrations.
5. Refer to Appendix 5 for a more comprehensive air change calculation procedure.

Table 6

Rating and Temperature Swing Multiplier (RSM) *com* *Res*

Method Used to Select Equipment	Summer Design	Temp. Swing 4.5	Temp. Swing 3.0
Selection Made at the Actual Summer Design Condition Using Manufacturer's Performance Data		0.90	1.00
Selection Made at the ARI Standard Rating Design Condition	85-90	0.85	0.95
	95	0.90	1.00
	100	0.95	1.05
	105	1.00	1.10
	110	1.05	1.15

Use this table in conjunction with calculation Procedure D.

Table 7A

Duct Loss Multipliers

	Duct Loss Multipliers	
Case I · Supply Air Temperatures Below 120°F *High eff 90% & Heat Pumps* Duct Location and Insulation Value	Winter Design Below 15°F	Winter Design Above 15°F
Exposed to Outdoor Ambient		
Attic, Garage, Exterior Wall, Open Crawl Space - None	.30	.25
Attic, Garage, Exterior Wall, Open Crawl Space - R2	.20	.15
Attic, Garage, Exterior Wall, Open Crawl Space - R4	.15	.10
Attic, Garage, Exterior Wall, Open Crawl Space - R6	.10	.05
Enclosed In Unheated Space		
Vented or Unvented Crawl Space or Basement - None	.20	.15
Vented or Unvented Crawl Space or Basement - R2	.15	.10
Vented or Unvented Crawl Space or Basement - R4	.10	.05
Vented or Unvented Crawl Space or Basement - R6	.05	.00
Duct Buried In or Under Concrete Slab		
No Edge Insulation	.25	.20
Edge Insulation R Value = 3 to 4	.15	.10
Edge Insulation R Value = 5 to 7	.10	.05
Edge Insulation R Value = 7 to 9	.05	.00
Case II · Supply Air Temperatures Above 120°F *Low eff 80%* Duct Location and Insulation Value	Winter Design Below 15°F	Winter Design Above 15°F
Exposed to Outdoor Ambient		
Attic, Garage, Exterior Wall, Open Crawl Space - None	.35	.30
Attic, Garage, Exterior Wall, Open Crawl Space - R2	.25	.20
Attic, Garage, Exterior Wall, Open Crawl Space - R4	.20	.15
Attic, Garage, Exterior Wall, Open Crawl Space - R6	.15	.10
Enclosed In Unheated Space		
Vented or Unvented Crawl Space or Basement - None	.25	.20
Vented or Unvented Crawl Space or Basement - R2	.20	.15
Vented or Unvented Crawl Space or Basement - R4	.15	.10
Vented or Unvented Crawl Space or Basement - R6	.10	.05
Duct Buried In or Under Concrete Slab		
No Edge Insulation	.30	.25
Edge Insulation R Value = 3 to 4	.20	.15
Edge Insulation R Value = 5 to 7	.15	.10
Edge Insulation R Value = 7 to 9	.10	.05

Table 7B
Duct Gain Multipliers

Duct Location and Insulation Value	Duct Gain Multiplier
Exposed to Outdoor Ambient	
Attic, Garage, Exterior Wall, Open Crawl Space - None	.30
Attic, Garage, Exterior Wall, Open Crawl Space - R2	.20
Attic, Garage, Exterior Wall, Open Crawl Space - R4	.15
Attic, Garage, Exterior Wall, Open Crawl Space - R6	.10
Enclosed In Unconditioned Space	
Vented or Unvented Crawl Space or Basement - None	.15
Vented or Unvented Crawl Space or Basement - R2	.10
Vented or Unvented Crawl Space or Basement - R4	.05
Vented or Unvented Crawl Space or Basement - R6	.00
Duct Buried In or Under Concrete Slab	
No Edge Insulation	.10
Edge Insulation R Value = 3 to 4	.05
Edge Insulation R Value = 5 to 7	.00
Edge Insulation R Value = 7 to 9	.00

Table 8
Shaded Glass Area

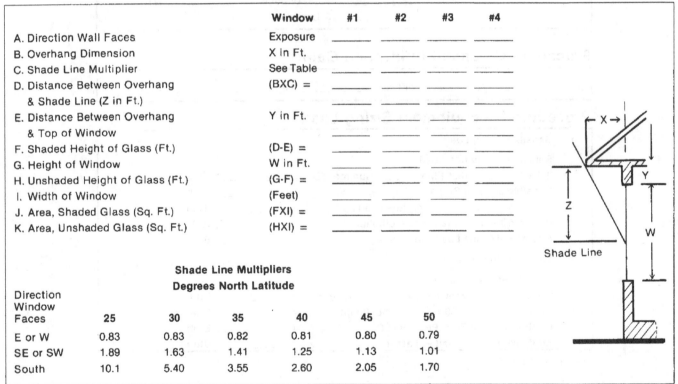

	Window	#1	#2	#3	#4
A. Direction Wall Faces	Exposure				
B. Overhang Dimension	X In Ft.				
C. Shade Line Multiplier	See Table				
D. Distance Between Overhang & Shade Line (Z in Ft.)	(BXC) =				
E. Distance Between Overhang & Top of Window	Y in Ft.				
F. Shaded Height of Glass (Ft.)	(D-E) =				
G. Height of Window	W in Ft.				
H. Unshaded Height of Glass (Ft.)	(G-F) =				
I. Width of Window	(Feet)				
J. Area, Shaded Glass (Sq. Ft.)	(FXI) =				
K. Area, Unshaded Glass (Sq. Ft.)	(HXI) =				

Shade Line Multipliers
Degrees North Latitude

Direction Window Faces	25	30	35	40	45	50
E or W	0.83	0.83	0.82	0.81	0.80	0.79
SE or SW	1.89	1.63	1.41	1.25	1.13	1.01
South	10.1	5.40	3.55	2.60	2.05	1.70

Use Table 8 to determine the square feet of shaded and unshaded glass areas beneath permanent external shading devices such as roof overhangs. A separate calculation is required for each window.

The total heat gain for a window that is partially shaded is equal to the sum of the heat gain through the shaded and unshaded areas. The heat gain through the shaded area is equal to the shaded area multiplied by the Table 3 heat transfer multiplier (HTM) for north (or external shading) glass. The heat gain through the unshaded area is equal to the unshaded area multiplied by the Table 3 heat transfer multiplier (HTM) for glass that has the same exposure as the window that is under consideration.

Shaded areas are not calculated for windows that face north, north-east or north-west because these exposures are not subjected to direct sunlight.

Refer to Section 5-2 for an example calculation.

Calculation Procedures A,B,C,D

Procedure A - Winter Infiltration HTM Calculation

1. Winter Infiltration CFM

   _____ AC/HR x _____ Cu. FT. x 0.0167 = _____ CFM

   Volume

2. Winter Infiltration Btuh

   1.1 x _____ CFM x _____ Winter TD = _____ Btuh

3. Winter Infiltration HTM

   _____ Btuh ÷ _____ Total Window = _____ HTM

   & Door Area

Procedure B - Summer Infiltration HTM Calculation

1. Summer Infiltration CFM

   _____ AC/HR x _____ Cu. FT. x 0.0167 = _____ CFM

   Volume

2. Summer Infiltration Btuh

   1.1 x _____ CFM x _____ Summer TD = _____ Btuh

3. Summer Infiltration HTM

   _____ Btuh ÷ _____ Total Window = _____ HTM

   & Door Area

Procedure C - Latent Infiltration Gain

0.68 x _____ gr. diff. x _____ CFM = _____ Btuh

Procedure D - Equipment Sizing Loads

1. Sensible Sizing Load

   Sensible Ventilation Load

1.1 x ____ Vent. CFM x ____ Summer TD	=	_____ Btuh
Sensible Load for Structure (Line 19)	+	_____ Btuh
Sum of Ventilation and Structure Loads	=	_____ Btuh
Rating & Temperature Swing Multiplier*	x	_____ RSM
Equipment Sizing Load - Sensible	=	_____ Btuh

2. Latent Sizing Load

   Latent Ventilation Load

0.68 x _____ Vent. CFM x _____ gr. diff.	=	_____ Btuh
Internal Loads = 230 x _____ No. People	+	_____ Btuh
Infiltration Load From Procedure C	+	_____ Btuh
Equipment Sizing Load — Latent	=	_____ Btuh

* Refer to Table 6

Table 9

Requirements For Mechanical Ventilation

1. **General Ventilation**

   If the Table 5 Summer Infiltration Calculation indicates that the summer infiltration rate is less than 0.35 ac/hr, mechanical ventilation may be required. When mechanical ventilation is indicated, provide 50 CFM of outdoor air or the quantity of outdoor air CFM that is indicated by the following formula: (whichever is greater).

   $$\text{CFM} = \frac{(0.35 - \text{Summer ac/hr})}{60} \times \text{cu. ft. of Conditioned Space}$$

2. **Combustion Air**

   When fossil fuel heating equipment obtains combustion air from within the building envelope, the Table 5 Winter Infiltration Calculation should indicate that at least 0.50 CFM of outdoor air per 1000 Btuh of input capacity is available for combustion air. If this is not the case, mechanical ventilation may be required. Also refer to Section 5 of ANSI Z223.1/NFPA 54 - 1984.

3. **Exhaust Systems**

   Residential exhaust systems are intended to operate intermittently. Therefore, each exhaust system should be equipped with a readily accessible switch or other means for shut off. Dampers should be installed for the purpose of isolating exhaust openings when the exhaust fans are not operating. Typical exhaust CFM for kitchens and baths are:

Kitchens - 100 CFM	Each Bath - 50 CFM

Table 10
R-values of Common Building Materials

(Used with permission of The American Society of Heating, Refrigerating and Air-Conditioning Engineers)

Position of Air Space	Direction of Heat Flow	Air Space Thickness, in.	Thermal Resistance (R)		
			Bright Aluminum Foil	Aluminum Painted Paper	Non-Reflective
No. 1 Still Air Surfaces					
(a) Horizontal	Up		1.32	1.10	0.61
(b) 45° Slope	Up		1.37	1.14	0.62
(c) Vertical	Horiz.		1.70	1.35	0.68
(d) 45° Slope	Down		2.22	1.67	0.76
(c) Horizontal	Down		4.55	2.70	0.92
No. 2 Air Spaces					
(a) Horizontal	Up (Winter)	¾ to 4	2.06	1.62	0.85
(b) Horizontal	Up (Summer)	¾ to 4	2.75	1.87	0.80
(c) 45° Slope	Up (Winter)	¾ to 4	2.22	1.71	0.88
(d) Vertical	Horiz. (Winter)	¾ to 4	2.62	1.94	0.94
(e) Vertical	Horiz. (Summer)	¾ to 4	3.44	2.16	0.91
(f) 45° Slope	Down (Summer)	¾ to 4	4.36	2.50	0.90
(g) Horizontal	Down (Winter)	¾	3.55	2.39	1.02
(h) Horizontal	Down (Summer)	¾	3.25	2.08	0.84
(i) Horizontal	Down (Winter)	1½	5.74	3.21	1.14
(j) Horizontal	Down (Summer)	1½	5.24	2.76	0.93
(k) Horizontal	Down (Winter	4	8.94	4.02	1.23
(l) Horizontal	Down (Summer)	4	8.08	3.38	0.99
No. 3 Moving Air Surfaces (Any Position or Direction)					
(a) 15 mph Wind (Winter)					0.17
(b) 7½ mph Wind (Summer)					0.25

Description	Density lb. per cu. ft.	Thermal Resistance (R)	
		Per Inch of Thickness	For Thickness Listed
No. 4 Building Board, Boards, Panels, Sheathing, etc.			
(a) Asbestos-cement board	120	0.25	
(b) Asbestos-cement board — ⅛ in.	120		0.03
(c) Gypsum or plaster board — ⅜ in.	50		0.32
(d) Gypsum or plaster board — ½ in.	50		0.45
(e) Plywood	34	1.25	
(f) Plywood — ¼ in.	34		0.31
(g) Plywood — ⅜ in.	34		0.47
(h) Plywood — ½ in.	34		0.63
(i) Plywood — ⅝ in.	34		0.78
(j) Plywood or wood panels — ¾ in.			0.94
(k) Wood fiber board, laminated or homogeneous	26	2.38	
	31	2.00	
	33	1.82	
(l) Wood fiber, hardboard type	65	0.72	
(m) Wood fiber, hardboard type — ¼ in.	65		0.18
(n) Wood, fir or pine sheathing — 25/32 in.			0.98
(o) Wood, fir or pine — 1⅝ in.			2.03

Table 10 (Continued)

Description	Density lb. per cu. ft.	Thermal Resistance (R)	
		Per Inch of Thickness	For Thickness Listed
No. 5 Building Paper			
(a) Vapor-permeable felt			0.06
(b) Vapor-seal, two layers or mopped 15 lb. felt			0.12
(c) Vapor-seal, plastic film			Negl
No. 6 Flooring Materials			
(a) Carpet and fibrous pad			2.08
(b) Carpet and rubber pad			1.23
(c) Cork tile ⅛ in.			0.28
(d) Floor tile or lineoleum — average value — ⅛ in.			0.05
(e) Terrazzo — 1 in.			0.08
(f) Wood subfloor — 25/32 in.			0.98
(g) Wood, hardwood finish — ¾ in.			0.68
No. 7 Insulating Materials, Blanket and Batt			
(a) Cotton fiber	0.8-2.0	3.85	
(b) Mineral wool, fibrous form, (2"-2¾") processed from rock, slag (3"-3½") or glass (3½"-3⅝") (5¼"-6½") (6"-7")	1.5-4.0		7 11 13 19 23
(c) Wood fiber	3.2-3.6	4.0	
No. 8 Insulating Materials, Board			
(a) Glass fiber	9.5-11.0	4.00	
(b) Wood or cane fiber Acoustical tile — ½ in.			1.19
(c) Wood or cane fiber Acoustical tile — ¾ in.			1.78
(d) Wood or cane fiber Interior finish, (plank, tile, lath)	15.0	2.86	
(e) Wood or cane fiber Interior finish, (plank, tile, lath) — ½ in.	15.0		1.43
(f) Roof deck slab, approximately — 1½ in.			4.17
(g) Roof deck slab, approximately — 2 in.			5.56
(h) Roof deck slab, approximately — 3 in.			8.33
(i) Sheathing, impregnated or coated	20.0	2.63	
(j) Sheathing, impregnated or coated — ½ in.	20.0		1.32
(k) Sheathing, impregnated or coated — 25/32 in.	20.0		2.06
No. 9 Insulating Materials, Board and Slabs			
(a) Cellular glass	9.0	2.50	
(b) Expanded Urethane		5.88	
(c) Expanded rubber	4.5	4.55	
(d) Hog hair, with asphalt binder	8.5	3.00	
(e) Expanded Polystyrene, molded beads	1.0	3.57	
(f) Expanded Polystyrene extruded	2.2	5.00	
(g) Wood shredded, cemented in preformed slabs	22.0	1.82	
(h) Mineral wool with resin binder	15.0	3.45	
(i) Mineral wool with asphalt binder	15.0	3.22	
No. 10 Insulating Materials, Loose Fill			
(a) Macerated paper or pulp products	2.5-3.5	3.57	
(b) Mineral wool, glass, slag, or rock	2.0-5.0	3.33	
(c) Sawdust or shavings	8.0-15.0	2.22	
(d) Silica Aerogel	7.6	5.88	
(e) Vermiculite, expanded	7.0-8.2	2.08	
(f) Wood fiber, redwood, hemlock, or fir	2.0-3.5	3.33	
(g) Wood fiber, redwood bark	3.0	3.22	
(h) Wood fiber, redwood bark	4.0	3.57	
(i) Wood fiber, redwood bark	4.5	3.84	

Table 10 (Continued)

Description	Density lb. per cu. ft.	Thermal Resistance (R)	
		Per Inch of Thickness	For Thickness Listed
No. 11 Roof Insulation. Preformed. for Use Above Deck			
(a) Approximately — ½ in.			1.39
(b) Approximately — 1 in.			2.78
(c) Approximately — 1½ in.			4.17
(d) Approximately — 2 in.			5.56
(c) Approximately — 2½ in.			6.67
(f) Approximately — 3 in.			8.33
No. 12 Masonry Materials. Concretes			
(a) Cement mortar	116	0.20	
(b) Gypsum-fiber concrete. 7½% gypsum, 12½% wood chips	51	0.60	
(c) Lightweight aggregates including	120	0.19	
	100	0.28	
expanded shale, clay or slate; expanded	80	0.40	
slags; cinders; pumice; perlite; vermiculite;	60	0.59	
also cellular concretes	40	0.86	
	30	1.11	
	20	1.43	
(d) Sand and gravel or stone aggregate, oven dried	140	0.11	
(e) Sand and gravel or stone aggregate, not dried	140	0.08	
(f) Stucco	116	0.20	
No. 13 Plastering Materials			
(a) Cement plaster, sand aggregate	116	0.20	
(b) Cement plaster, sand aggregate — ½ in.			0.10
(c) Cement plaster. sand aggregate — ¾ in.			0.15
(d) Gypsum plaster, lightweight aggregate — ½ in.	45		0.32
(e) Gypsum plaster, lightweight aggregate — ⅝ in.	45		0.39
(f) Gypsum plaster, lightweight aggregate on metal lath — ¾ in.			0.47
(g) Gypsum plaster, perlite aggregate	45	0.67	
(h) Gypsum plaster, sand aggretate	105	0.18	
(i) Gypsum plaster, sand aggregate — ½ in.	105		0.09
(j) Gypsum plaster, sand aggregate — ⅝ in.	105		0.11
(k) Gypsum plaster, sand aggregate on metal lath — ¾ in.			0.10
(l) Gypsum plaster, sand aggregate on wood lath			0.40
(m) Gypsum plaster, vermicalite aggregate	45	0.59	
No. 14 Masonry Units			
(a) Brick, common	120	0.20	
(b) Brick, face	130	0.11	
(c) Hollow clay tile, one cell deep — 3 in.			0.80
(d) Hollow clay tile, one cell deep — 4 in.			1.11
(e) Hollow clay tile. two cells deep — 6 in.			1.52
(f) Hollow clay tile. two cells deep — 8 in.			1.85
(g) Hollow clay tile. two cells deep — 10 in.			2.22
(h) Hollow clay tile, three cells deep — 12 in.			2.50
(i) Stone. lime or sand		0.08	
(j) Gypsum partition tile, 3 in. x 12 in. x 30 in. — solid			1.26
(k) Gypsum partition tile, 3 in. x 12 in. x 30 in. — 4 cell			1.35
(l) Gypsum partition tile. 4 in. x 12 in. x 30 in. — 3 cell			1.67

Table 10 (Continued)

Description	Density lb. per cu. ft.	Thermal Resistance (R)	
		Per Inch of Thickness	For Thickness Listed
No. 15 Concrete Blocks			
(a) Sand and gravel aggregate, three oval core — 4 in.			0.71
(b) Same as (15a) but — 8 in.			1.11
(c) Same as (15a) but — 12 in.			1.28
(d) Cinder aggregate, three oval core — 3 in.			0.86
(e) Same as (15d) but — 4 in.			1.11
(f) Same as (15d) but — 8 in.			1.72
(g) Same as (15d) but — 12 in.			1.89
(h) Sand and gravel aggregate, two core — 8 in., 36 lb.			1.04
(i) Same as (15h) but — with filled cores			1.93
(j) Lightweight aggregate, expanded shale, clay, slate or slag; pumice — 3 in.			1.27
(K) Same as (15j) but — 4 in.			1.50
(l) Same as (15j) but — 8 in.			2.00
(m) Same as (15j) but — 12 in.			2.27
(n) Lightweight aggregate, expanded shale, clay, slate or slag, pumice — 2 core, 8 in., 24 lb.			2.18
(o) Same as (15n) but — with filled cores			5.03
(p) Same as (15n) but — 3 core, 6 in., 19 lb.			1.65
(q) Same as (15p) but — with filled cores			2.99
(r) Same as (15n) but — 3 core, 12 in., 38 lb.			2.48
(s) Same as (15r) but — with filled cores			5.82
No. 16 Roofing			
(a) Asbestos-cement shingles	120		0.21
(b) Asphalt roll roofing	70		0.15
(c) Asphalt shingles	70		0.44
(d) Built-up roofing — 3/8 in.	70		0.33
(e) Slate — 1/2 in.			0.05
(f) Sheet Metal		Negl	
(g) Wood shingles			0.94
No. 17 Siding Materials (On Flat Surface)			
(a) Wood shingles, 16 in., 7½ in. exposure			0.87
(b) Wood shingles, double, 16 in., 12 in. exposure			1.19
(c) Wood shingles, plus insulation 5/16 in. backer board			1.40
(d) Asbestos-cement siding, ¼ in., lapped or shingles			0.21
(e) Asphalt roll siding			0.15
(f) Asphalt insulating siding (½ in. board)			1.46
(g) Wood siding drop, 1 in. x 8 in.			0.79
(h) Wood siding, bevel, ½ in. x 8 in., lapped			0.81
(i) Wood siding, bevel, ¾ in. x 10 in., lapped			1.05
(j) Wood siding, plywood, 3/8 in., lapped			0.59
(k) Structural glass			0.10
No. 18 Woods			
(a) Maple, oak, and similar hardwoods	45	0.91	
(b) Fir, pine, and similar soft woods	32	1.25	
(c) Fir, pine, and similar soft woods — 25/32 in.	32		0.98
(d) Fir, pine and similar soft woods — 1⅝ in.	32		2.03
(e) Fir, pine, and similar soft woods — 2⅝ in.	32		3.28
(f) Fir, pine, and similar soft woods — 3⅝ in.	32		4.55

Appendix A-1
Mobile Home Load Calculations

The methods and procedures which are outlined in Manual J can be used to calculate the heating and cooling loads for mobile homes. The J-1 Form, and the Calculation Procedures A,B,C, and D can be used to calculate the loads that are associated with mobile home construction. But, the wall and roof HTM values listed in Tables 2 and 4 of Manual J do not apply to mobile homes.

A1-1 Wall, Roof and Floor HTM (Heating)
Table A1-1 (in this section) contains the mobile home HTM values for heating. Use Table A1-1 to calculate the heat loss through metal walls, roofs and floors. The HTM values in this table were calculated by multiplying the winter T.D. by the metal wall, roof or floor U value.

A1-2 Wall, Roof and Floor ETDs (Cooling)
One of the biggest differences between a mobile home and a conventional structure is that the mobile home is a much lighter structure and it has less thermal mass associated with the walls and roof. This light weight insulated metal construction has less ability to absorb and hold heat than the heavier walls used for conventional homes. Even if both types of construction have the same "R" (thermal resistance) value, the lighter weight wall will have a higher "HTM" (heat transfer multiplier).

Table A1-2 contains the mobile home ETD values for walls, roofs and floors. This table is based on ETD information contained in the ASHRAE GRP-158 Load Calculation Manual. The ETD's listed in this table are average values. (Refer to Section 7-15 for a discussion of why average values are used.)

A1-3 Roof, Wall, and Floor HTM (Cooling)
The values in Table A1-2 can be multiplied by the wall, roof or floor transmission coefficient (U value) to find the equivalent Manual J HTM value. Table A1-3 contains the results of this calculation. Use Table A1-3 when calculating the loads for mobile home walls, roofs and floors.

A1-4 Mobile Home Load Calculation Procedure
The procedure is basically the same as the procedures discussed in Sections 1 through 6 of this manual. However, the HTM values listed in Tables A1-1 and A1-3 are used in place of the corresponding data listed for construction numbers 12, 18 and 20 in Tables 2 and 4.

Design T.D.	10		15	20	25	30		35
Daily Range	L	M	LMH	LMH	LMH	M	H	H
Wall ETD	30		35	40	45	50		55
Roof ETD	60		65	70	75	80		85
Floor ETD	10		15	20	25	30		35

Table A1-2
ETD Table for Metal Walls and Roofs

Winter T.D.	20	25	30	35	40	45	50	55	60	65	70	75	80	85	90
Metal Wall															
1" INS (R = 4)	5.00	6.25	7.50	8.75	10.00	11.25	12.50	13.75	15.00	16.25	17.50	18.75	20.00	21.25	22.50
2" INS (R = 7)	2.86	3.57	4.29	5.00	5.71	6.43	7.14	7.86	8.57	9.29	10.00	10.71	11.43	12.14	12.86
3" INS (R = 10)	2.00	2.50	3.00	3.50	4.00	4.50	5.00	5.50	6.00	6.50	7.00	7.50	8.00	8.50	9.00
Metal Roof															
1" INS (R = 4)	5.00	6.25	7.50	8.75	10.00	11.25	12.50	13.75	15.00	16.25	17.50	18.75	20.00	21.25	22.50
2" INS (R = 7)	2.86	3.57	4.29	5.00	5.71	6.43	7.14	7.86	8.57	9.29	10.00	10.71	11.43	12.14	12.86
3" INS (R = 10)	2.00	2.50	3.00	3.50	4.00	4.50	5.00	5.50	6.00	6.50	7.00	7.50	8.00	8.50	9.00
Carpeted Floor															
No Insulation	10.00	12.50	15.00	17.50	20.00	22.50	25.00	27.50	30.00	32.50	35.00	37.50	40.00	42.50	45.00
1" INS (R = 4)	5.00	6.25	7.50	8.75	10.00	11.25	12.50	13.75	15.00	16.25	17.50	18.75	20.00	21.25	22.50
2" INS (R = 7)	2.86	3.57	4.29	5.00	5.71	6.43	7.14	7.86	8.57	9.29	10.00	10.71	11.43	12.14	12.86

Table A1-1 HTM Values for Mobile Homes (Heating)

Summer Design T.D. Daily Range	10 L M H	15 L M H	20 L M H	25 L M H	30 L M H	35 L M H
Metal Wall — Dark						
1" INS (R = 4)	7.50	8.75	10.00	11.25	12.50	13.75
2" INS (R = 7)	4.29	5.00	5.71	6.43	7.14	7.86
3" INS (R = 10)	3.00	3.50	4.00	4.50	5.00	5.50
Metal Wall — Light						
1" INS (R = 4)	4.88	5.69	6.50	7.31	8.13	8.94
2" INS (R = 7)	2.79	3.25	3.71	4.18	4.64	5.11
3" INS (R = 10)	1.95	2.28	2.60	2.93	3.25	3.58
Metal Roof — Dark						
1" INS (R = 4)	15.00	16.25	17.50	18.75	20.00	21.25
2" INS (R = 7)	8.57	9.29	10.00	10.71	11.43	12.14
3" INS (R = 10)	6.00	6.50	7.00	7.50	8.00	8.50
Metal Roof — Light						
1" INS (R = 4)	9.75	10.56	11.38	12.19	13.00	13.81
2" INS (R = 7)	5.57	6.04	6.50	6.96	7.43	7.89
3" INS (R = 10)	3.90	4.23	4.55	4.88	5.20	5.53
Carpeted Floor						
No Insulation	5.00	7.50	10.00	12.50	15.00	17.50
1" INS (R = 4)	2.50	3.75	5.00	6.25	7.50	8.75
2" INS (R = 7)	1.43	2.14	2.86	3.57	4.29	5.00

Table A1-3 HTM Values for Mobile Homes (Cooling)

Appendix A-2
Multi-Zone Systems

Until recently, multi-zone heating and cooling equipment was not commonly available for residential applications. However, this situation is changing and these systems have the potential to become an important factor in the residential market. Therefore, it is necessary for the industry to address the technical questions that are associated with designing these systems. With this purpose in mind, ACCA is offering the following procedure as a first step in developing an industry standard which will specify the calculations and procedures that should be applied to residential multi-zone equipment. Interested individuals, societies, associations or manufacturers are encouraged to comment on the procedure which is outlined in this section.

Multi-Zone System Performance Characteristics:

- All zones isolated by a full partition.
- The temperature in each zone can be adjusted independently of the other zones.
- The thermostat setting in zones which are unoccupied can be set back in the heating season or set up in the cooling season while design conditions are maintained in the occupied zones.
- When climatic conditions allow, the cooling equipment in any zone or any number of zones can be turned off while design conditions or set up conditions are maintained in the rest of the structure.
- Equipment size (and installed KW) can be reduced when the output capacity is based on load estimates which recognize that the rooms in the structure (which are unoccupied) need not be maintained at the indoor design condition.
- Annual energy consumption can be reduced when the equipment is sized and controlled to take advantage of diversity associated with heating and cooling loads or the reduction in operating hours associated with occupancy schedules.

A2-1 Multi-Zone Equipment
Multi-zone equipment may consist of equipment which controls flow of air to each zone or equipment that controls the flow of refrigerant to each zone. Equipment which controls air flow consists of a ducted variable air volume system that is characterized by a central air handler which contains the fan and a cooling coil. Equipment that controls refrigerant flow is characterized by a single central condensing unit which supplies refrigerant to multiple fan coils.

A2-3 Zone Cooling Loads
Cooling loads associated with a single room or with a few rooms are zone loads. The cooling load in any particular zone can be quite sensitive to the position of the sun. For example, zones that face to the east usually experience a maximum cooling load (peak load) during the morning and zones which face west usually peak during the late afternoon. The design cooling load (sensible) for a zone is typically calculated at the peak load condition.

A2-4 Block Cooling Loads
The block load is equal to the largest cooling load that can occur when a number of rooms or zones are considered as a group. The block load is usually less than the sum of the peak zone loads because the peak zone loads usually do not occur simultaneously; (referred to as "diversity"). For example, since the peak load in an east facing zone will usually not occur at the same time as the peak load in a west facing zone, the block load must be less than the sum of the two peak loads. The block load can be used as the design cooling load whenever a group of rooms provide the opportunity to take advantage of diversity.

A2-5 Manual J Sensible Load

The Manual J sensible load calculation is similar to a block load because the sensible cooling calculations are based on equivalent temperature difference and solar gain data that has been averaged over most of daylight hours. Note that the standard Manual J load estimate makes an allowance for a three degree temperature swing during peak load conditions.

A2-6 Zone Correction Factors

When a room (or a few rooms) has a limited number of exposures, the effects of diversity are minimized and the peak zone load provides a better approximation of the design condition. In this case, the peak zone load may exceed the calculated Manual J load if the room or zone has exposures that are sensitive to the position of sun. The multipliers listed in Table A2-1 can be used to estimate the peak sensible design load for a room or zone. To keep the zone calculation consistent with the standard Manual J calculation, the factors in Table A2-1 include an allowance for a 3 degree swing in indoor temperature during peak load conditions.

To use Table A2-1, multiply the standard Manual J sensible load estimate by the correction factor which corresponds to the primary exposure for the room or zone which is under consideration. Note that window area is the most important factor for selecting the primary exposure. If the room or zone has two exposures of equal importance (for example "west" and "south"), use the correction factor that corresponds to the intermediate compass point. In this case, it's "southwest."

Table A2-1
Room Sensible Load Correction Factors

Exposure	Glass Area as Percent of Wall Area				
	10%	15%	20%	25%	30%
North	1.00	1.00	1.00	1.00	1.00
Northeast	1.00	1.00	1.00	1.00	1.00
East	1.00	1.00	1.00	1.00	1.05
Southeast	1.00	1.05	1.05	1.10	1.10
South	1.15	1.20	1.25	1.30	1.35
Southwest	1.20	1.30	1.35	1.35	1.40
West	1.25	1.25	1.25	1.25	1.25
Northwest	1.15	1.15	1.20	1.20	1.20

A2-7 Peak Room or Zone Cooling Load

When multi-zone equipment is used, the peak sensible load for each room or zone is equal to the standard Manual J cooling load multiplied by a correction factor from Table A2-1. This load determines the room CFM or the capacity and size of the indoor fan coil.

A2-8 Room or Zone Cooling Loads at Set-Up

The sensible cooling load for rooms that are maintained at a temperature that exceeds the indoor summer design temperature is equal to the Manual J load calculated at the set-up temperature multiplied by the appropriate correction factor from Table A2-1.

A2-9 Central Equipment Cooling Loads (Sensible)

The design sensible load associated with a group of rooms which are not able to take advantage of diversity (characterized by having one or two primary exposures) is equal to the sum of peak sensible cooling loads of the various rooms. For this calculation, the peak sensible load in each room should be estimated at the indoor design temperature or at the set-up temperature that will exist within the room during the design condition. Include partition loads when conditioned rooms are adjacent to unconditioned rooms. Do not include loads for rooms which are not cooled during the design condition.

Note: Diversity applies when the multi-zone equipment serves the entire house (or an entire floor of a multi-story house in this case), and the design sensible load on the unit equals the standard Manual J (block) load for the rooms in questions.

A2-10 Design Latent Cooling Load

The design latent load on the central refrigeration equipment shall be based on the same combination of rooms and operating conditions which were used for the sensible design load estimate. Include the internal and infiltration latent loads that are associated with these rooms. Use calculation procedures B, C and D to estimate the total latent load associated with a group of rooms.

A2-11 Examples of Multi-Zone Applications

Figures A2-1 through A2-5 show how the design ing loads are affected by the manner in which the house is zoned.

Manual J Cooling Loads	Primary Exposure	Design Load @ 75°F	Set-Up Load @ 85°F
Living Room	N & W	1800	1350
Dining Room	West	4200	3150
Kitchen	S & W	5100	3825
Family Room	South	2200	1650
Master Bedroom	North	1800	1350
Bedroom 3	S & E	3300	2475
Bedroom 2	East	2500	1875
Bedroom 1	East	2500	1875
Bath 1	North	600	
Bath 2	North	400	
Design Latent Load =		3,440	

Glass area approximately 15% of wall space.

Figure A2-2 Floor Plan and the Manual J Cooling Loads for Example House

When a central ducted VAV system cools the entire house, all exposures are of equal importance and the total load on the central equipment can be estimated by summing the uncorrected Manual J room loads.

The design sensible cooling load is equal to 24,400 Btu/Hr. Notice that the total sensible load is reduced if the cooling loads for rooms which are usually unoccupied during the day time are calculated at the set up temperature.

The correction factors are used to estimate the room loads and to determine the room CFM.

The design latent load is 3,440 Btuh and is equal to the latent load that was calculated for the entire house by the standard Manual J procedure.

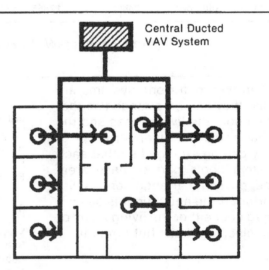

Central Ducted VAV System

	Primary Exposure	Manual J Load @ 75°F	Room Load Factor	Room Design Load	Manual J Load @ 75°F	Manual J Load @ 85°F	Liv @ 75°F Bath @ 75°F SLP @ 85°F
Living Room	N & W	1800	1.15	2070	1800	NA	1800
Dining Room	West	4200	1.25	5250	4200	NA	4200
Kitchen	S & W	5100	1.30	6630	5100	NA	5100
Family Room	South	2200	1.20	2640	2200	NA	2200
Master Bedroom	North	1800	1.00	1800	NA	1350	1350
Bedroom 3	S & E	3300	1.05	3465	NA	2475	2475
Bedroom 2	East	2500	1.00	2500	NA	1875	1875
Bedroom 1	East	2500	1.00	2500	NA	1875	1875
Bath 1	North	600	1.00	600	600	NA	600
Bath 2	North	400	1.00	400	400	NA	400
Total Sensible Load =		24400		Unoccupied Rooms at 85°F =			21875

*Glass area equals 15% of wall area.

Figure A2-3 Central Ducted VAV System Cools Entire House

When a single multi-zone system is sized to cool either the living areas or the sleeping areas (but not both simultaneously) the design sensible cooling load is equal to 17,590 Btu/Hr. Note that the room load correction factors are required because the living areas and the sleeping areas are associated with specific exposures.

The design latent load is equal to the latent load that is associated with the living areas which equals 2880 Btuh (1040 for infiltration and 1840 for occupants).

Central Equipment Ducted VAV or Multiple DX Coil System

	Primary Exposure	Manual J Load @ 75 F	Zone Load Factor	Adjst'd Load @ Design	Liv @ 75 Bath @ 75 SLP-Off	Liv-Off SLP @ 75 Bath @ 75
Living Room	N & W	1800	1.15	2070	2070	0
Dining Room	West	4200	1.25	5250	5250	0
Kitchen	S & W	5100	1.30	6630	6630	0
Family Room	South	2200	1.20	2640	2640	0
Master Bedroom	North	1800	1.00	1800	0	1800
Bedroom 3	S & E	3300	1.05	3465	0	3465
Bedroom 2	East	2500	1.00	2500	0	2500
Bedroom 1	East	2500	1.00	2500	0	2500
Bath 1	North	600	1.00	600	600	600
Bath 2	North	400	1.00	400	400	400
Total Sensible Loads =		24400		27855	17590	11265

SLEEPING AREAS
75 Degrees or Off

LIVING AREAS
75 Degrees or Off

Figure A2-4 Single Multi-Zone System Sized to Cool Part of the House

When two multi-zone systems are used they can be arranged so that one cools the living areas and the other cools the sleeping areas or they can be arranged so that each system cools part of each area. The physical arrangement is important if the equipment is designed to cool either the living areas or the sleeping areas but not both.

Unit A	Primary Exposure	Manual J Load @ 75°F	Zone Load Factor	Adjst'd Load @ Design	Manual J Latent Load
Living Room	N & W	1800	1.15	2070	Infil't
Dining Room	West	4200	1.25	5250	Plus
Kitchen	S & W	5100	1.30	6630	Eight
Family Room	South	2200	1.20	2640	People
Total Loads =		13300		16590	2880
Unit B					
Master Bedroom	North	1800	1.00	1800	
Bedroom 3	S & E	3300	1.05	3465	Infil't
Bedroom 2	East	2500	1.00	2500	Plus
Bedroom 1	East	2500	1.00	2500	Eight
Bath 1	North	600	1.00	600	People
Bath 2	North	400	1.00	400	
Total Loads =		11100		11265	2400

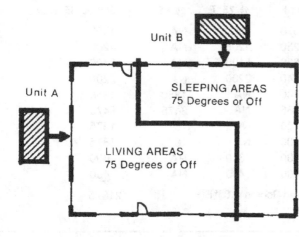

Unit B

Unit A

SLEEPING AREAS
75 Degrees or Off

LIVING AREAS
75 Degrees or Off

Units "A" and "B" are arranged so that unit "A" cools the living areas and unit "B" cools the sleeping areas. In this case the total sensible capacity is equal to the sum "A" and "B" capacities: 16590 + 11265 = 27855 Btuh.

Compare the total design load 27855 Btuh for this arrangement with the design load 17590 Btuh for the arrangement shown by Figure A2-6.

Figure A2-5 Two Multi-Zone Units Arranged to Cool Living or Sleeping Areas

When two multi-zone systems are used they can be arranged so that one cools the living areas and the other cools the sleeping areas or they can be arranged so that each system cools part of each area. The physical arrangement is important, refer to Figure A2-5 for an alternate arrangement.

Units "C" and "D" are arranged so that each unit cools both the living and sleeping areas. In this case the total sensible capacity is equal to the sum "C" and "D" capacities:
8320 + 9270 = 17590 Btuh.

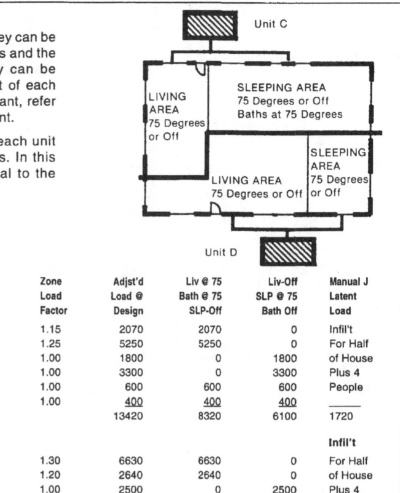

Unit C	Primary Exposure	Manual J Load @ 75° F	Zone Load Factor	Adjst'd Load @ Design	Liv @ 75 Bath @ 75 SLP-Off	Liv-Off SLP @ 75 Bath Off	Manual J Latent Load
Living Room	N & W	1800	1.15	2070	2070	0	Infil't
Dining Room	West	4200	1.25	5250	5250	0	For Half
Master Bedroom	North	1800	1.00	1800	0	1800	of House
Bedroom 1	East	3300	1.00	3300	0	3300	Plus 4
Bath 1	North	600	1.00	600	600	600	People
Bath 2	North	400	1.00	400	400	400	
	Total Loads =12100			13420	8320	6100	1720
Unit D							**Infil't**
Kitchen	S & W	5100	1.30	6630	6630	0	For Half
Family Room	South	2200	1.20	2640	2640	0	of House
Bedroom 2	East	2500	1.00	2500	0	2500	Plus 4
Bedroom 3	S & E	2500	1.05	2625	0	2625	People
	Total Loads = 12300			14395	9270	5125	1720

Figure A2-6 Two Multi-Zone Units, Each Cools Both the Living and Sleeping Areas

A2-12 Design Heating Loads
The heating loads associated with a room or with a few rooms are zone loads. Room or zone heating loads are not senstive to the direction the room or zone faces. The individual room loads are equal to the Manual J loads as calculated by the standard Manual J procedure. The total zone heating load for a group of rooms is equal to the sum of the heating loads that are associated with the individual rooms.

A2-13 Setback Heating Loads
The standard Manual J procedure can be used to estimate setback heating loads if the room load calculations are made at the setback temperature instead of the indoor design temperature.

A2-14 Equipment Output Heating Capacity
The heating capacity of multi-zone equipment should be based on heating loads which are calculated at the winter design condition.

Appendix A-3
Multi-Family Structures

Many multi-family structures are characterized by dwelling units that have only one or two walls which are exposed to the sun. In addition, these units may or may not have an exposed roof or exposed floor, and they may be affected by the temperature in the adjacent unit or units.

A3-1 Basic Cooling Loads
Use the standard Manual J calculation procedure to calculate the sensible and latent cooling loads. When performimg this calculation, use the infiltration evaluation table provided in this appendix to estimate the infiltration loads. Do not include any loads which may be associated with partitions, ceiling or floors which separate dwelling units. Assume all adjacent dwellings are maintained at the indoor design condition.

A3-2 Room Cooling Load Correction
Since the room cooling load is sensitive to the orientation of the exposed surfaces, the sensible cooling load calculation which is described above should be modified by the room load correction factors which appear in Appendix 3 and which are included in this paragraph.

If a room has exposed walls that are adjacent to each other, the exposed glass area is the most important factor for selecting the primary exposure. If one wall has a much larger glass area than the other, use the correction factor which corresponds to the direction in which that wall faces. If both walls have glass area which are approximately equal, use the correction factor which corresponds to the intermediate compass point.

A3-3 Central Equipment Cooling Load
The design sensible load for the entire dwelling is equal to the sum of the corrected room loads. The design latent load for the entire dwelling is equal to the latent load that is calculated by the standard Manual J procedure.

A3-4 Infiltration Loads
Because these types of dwelling units have less exposed wall and glass area than a single family structure, the infiltration loads will be somewhat lower. Use the table below to estimate the infiltration load for this type of construction, or use the procedure which is outlined in Appendix 5.

A3-5 Heating Loads
Use the standard Manual J calculation procedure to calculate the heating loads for each room and for the entire dwelling. When performing this calculation, use the infiltration evaluation table provided in this appendix to estimate the infiltration loads. Do not include any loads which may be associated with partitions, ceilings or floors which separate dwelling units. Assume all adjacent dwellings are maintained at the indoor design condition.

A3-6 Partition Loads
Since a given dwelling unit could conceivably be surrounded by as many as four unoccupied and unconditioned spaces, some consideration should be given to adding a partition load and/or a ceiling or floor load to the central equipment cooling and heating loads. It is the designers option to include these loads unless a regulation or code specified otherwise.

Table A2-1
Room Sensible Load Correction Factors

| Exposure | Glass Area as Percent of Wall Area | | | | |
	10%	15%	20%	25%	30%
North	1.00	1.00	1.00	1.00	1.00
Northeast	1.00	1.00	1.00	1.00	1.00
East	1.00	1.00	1.00	1.00	1.05
Southeast	1.00	1.05	1.05	1.10	1.10
South	1.15	1.20	1.25	1.30	1.35
Southwest	1.20	1.30	1.35	1.35	1.40
West	1.25	1.25	1.25	1.25	1.25
Northwest	1.15	1.15	1.20	1.20	1.20

Winter Air Changes Per Hour

Floor Area	900 or less	900-1500	1500-2100	over 2100
Best	0.4	0.3	0.2	0.2
Average	1.0	0.8	0.6	0.5
Poor	1.8	1.3	0.9	0.8
For each fire place add:			Best Average Poor	
			0.1 0.2 0.6	

Summer Air Changes Per Hour

Floor Area	900 or less	900-1500	1500-2100	over 2100
Best	0.2	0.2	0.2	0.2
Average	0.5	0.4	0.3	0.3
Poor	0.8	0.6	0.5	0.4

Appendix A-4
Energy Consumption and Operating Cost

Energy consumption and operating cost estimates provide important design information which can be used to select heating and cooling equipment which is the most economical or which provides the best return on investment. In some states, annual energy budget estimates are required if the building envelope design and/or the HVAC system performance deviate from prescriptive energy code requirements.

A4-1 Types of Residential HVAC Systems
Each residential HVAC system has unique operating characteristics which must be considered when energy consumption and operating cost are estimated. Some of the more common residential systems are listed below.

- Gas and propane furnaces (with or without standing pilot).
- Oil and electric furnaces.
- Air source heat pumps (heating), with electric auxiliary heat.
- Air source heat pumps (heating), with gas, propane or oil furnace.
- Water source heat pumps (heating), with electric auxiliary heat.
- Add on water source heat pumps (heating) with gas, propane or oil furnace.
- Electric baseboard radiation.
- Air cooled air conditioner or air source heat pump (cooling).
- Water source heat pump (cooling).
- Water coil cooling (well water cooling).
- Multi-zone systems (refrigerant side zoned).
- Multi-zone systems (air side zoned).
- Multi-zone radiation systems (hydronic or electric resistance).
- Thermal storage systems.
- Solar and hybrid convention/solar systems.

A4-2 Energy Calculation Methods
Residential heating and cooling energy consumption and operating cost are commonly estimated by one of the following methods:

- Some form of degree day calculation.
- By equations which are based on heating or cooling load hour data and equipment efficiency (HSPF, SEER, AFUE) descriptors.
- By a "bin method" calculation.
- By a "hour by hour" calculation.

The degree day and the efficiency descriptor equations are popular because the calculations are simple to perform. However, these methods are not as accurate as a comprehensive bin calculation or an hourly calculation.

Bin and hourly calculations are considerably more complicated, they require high level of technical knowledge and mathematical skill and are usually performed on a computer. The accuracy of the bin and hourly calculations can also vary with the amount of detail and the accuracy of the assumptions that are included in the model.

The following sections illustrate how the various types of energy estimates are performed. These examples are only intended to illustrate the basic procedure; refer to the ASHRAE handbooks and documents published by the department of energy for a more detailed discussion of calculation procedures.

A4-3 Degree Day Method
The Degree Day Method is normally used to estimate the energy consumption of gas, oil, electric resistance, or coal heating systems. This method should not be directly applied to heat pumps, add-on heat pumps or zoned systems. However, there are procedures which can be used to relate bin hours and degree day weather data (for a particular location) so that the degree day format can be applied to heat pumps.

The Degree Day formula shown in this manual is presented in its simplest form without the corrections factors that usually appear with this equation. This was done because the accuracy of the Degree Day Method and the corresponding corrections factors have been subject to some controversy. Refer to the latest ASHRAE fundamentals volume for a complete discussion of the Degree Day Method.

Also note that the Degree Day Method is based on a reference temperature which is equal the outdoor temperature at which the building envelope just begins to require heat. Historically this reference temperature was 65°F and was based on the following assumptions:

- The house has minimum to average insulation levels.
- The thermostat is set at 70°F.
- The internal gains provide enough heat to maintain 70°F in the house when the outdoor temperature is 65°F (5 degree heating credit).

Structures which have above average insulation levels, passive solar features, above average internal loads or thermostat settings other than 70°F should not be evaluated at the 65°F reference temperature.

Degrees days are recorded by the weather bureau. For convenience, Table 1 tabulates Degree Day data (65°F base) for various locations. The Degree Days listed are for the entire heating season.

These values are estimated as follows:

When the outside temperature falls below 65°F, heat will be required to maintain the temperature within the house. The average outside temperature is estimated by adding the high and low temperatures for a given day and dividing by 2. (A high of 42°F and a low of 20°F would be equivalent to a 31°F average temperature.) Degree days are defined as the difference between the average temperature and the 65°F base temperature. Therefore, 65°F - 31°F equals 34 degree days.

$$\text{Degree days} = 65 - (high + low)/2$$

In its basic form, the degree day equation is used as follows:

$$F = \frac{HL \times 24 \times DD}{E \times P \times T.D.}$$

Where:
HL is the design heating load (Btuh).
DD is the degree days from Table 1.
24 accounts for 24 hours per day.
E is the seasonal efficiency of the equipment.
P is the heating value of the fuel.
T.D. is the design temperature difference (°F).
F is the annual fuel consumption.

FUEL	E	P	UNITS	(F) UNITS
Gas	50-95%	1,025	Btu/cu. ft.	cu. ft.
Propane	50-95%	91,500	Btu/gal.	gal.
Oil	50-90%	138,000	Btu/gal.	gal.
Electric Resistance	90-100%	3,413	Btu/kwh.	Kwh.
Coal	50-65%	13,000	Btu/lb.	lb.

Example: A house has a design heat loss of 65,000 Btuh. The design temperature difference is 70 - (-5) = 75°F. The house is located in Des Moines, Iowa, which experiences 6,610 degree days. The following table summarizes the results of the degree day calculation, assuming the seasonal efficiencies indicated by the table.

Degree Day Fuel Consumption Estimate

FUEL	E	P	ENERGY	UNITS
Gas	.60	1,025	223,600	cu. ft./yr.
Propane	.60	91,500	2,500	gal./yr.
Oil	.60	138,000	1,660	gal./yr.
Electricity	1.00	3,413	40,280	Kwh./yr.
Coal	.55	13,000	19,230	lb./yr.

A4-4 Efficiency Descriptors
Energy efficiency descriptors such as AFUE, HSPF, SEER, COP and EER are useful for comparing similar types of equipment which are applied to similar load and weather patterns. However, energy estimates that are based on these descriptors may be in error when they are applied to a specific structure in a specific location.

- Local weather patterns can be substantially different than the weather data that was used to calculate the descriptor values.
- Inability to account for variations in occupancy schedules, and equipment operating schedules.
- Inability to account for variations in heating and cooling loads that are related to solar and internal loads.
- Inability to account for variations in cooling equipment running times that are caused by differences in latent loads.
- The effect of on/off cycling on the overall equipment efficiency is not always based on actual test data. (Default value used for the cycling degradiation coefficient.)
- COP and EER ratings are steady state ratings. They provide no indication of part load efficiency and they do not consider the effect of weather or heating and cooling loads.

In addition, energy descriptors should not be used to compare dissimilar types of equipment:

- The AFUE rating for furnaces is not equivalent to the HSPF rating for heat pumps.
- The HSPF and SEER efficiency ratings that are used for air source heat pumps cannot be compared to the COP and EER ratings that are used for water source heat pumps.
- A single zone heat pump and a multi-zone heat pump can have identical HSPF and SEER ratings but they will consume different amounts of energy (on an annual basis) because the multi-zone equipment has the potential to take advantage of diversity in the space loads and occupancy schedules.
- COP and EER are the only descriptors that relate directly to the connected electrical load (KW). However, these descriptors do not indicate that the effective on line power draw required for multi-zone system can be less than that which is required for a single zone system.

A4-5 AFUE Rating
The Annual Fuel Efficiency Ratings (AFUE) for gas and oil fired heating equipment is listed in the "Directory of Certified Furnace and Boiler Efficiency Ratings" published by the Gas Appliance Manufacturers Associations (GAMA).

The AFUE rating was developed to account for the effects of on/off cycling, flue losses, and other transient and secondary factors that effect furnace efficiency under actual operating conditions.

The AFUE number can be used to estimate the annual heating energy that is consumed by the furnace by using the following equations:

$$\text{Burner Btu/yr.} = \frac{0.77 \times HLH \times QH}{AFUE}$$

$$\text{*Pilot Btu/yr.} = 8760 - \frac{0.77 \times HLH \times QH}{AFUE \times QB} \times QP$$

$$\text{Fan Kwh/yr.} = \frac{0.77 \times HLH \times QH}{AFUE \times QB} \times KW$$

Btu/yr = Annual energy input to the furnace. (Note that this does not include fan energy.)

Kwh/yr. = Annual energy to blower (intermittent operation.)

8760 = Hours in one year.

0.77 = A factor which is required to make the design heating load hours correlate with the effective heating load hours experienced under actual operating conditions.

HLH = Heating Load Hours from the map in the GAMA directory.

QH = The design heating load - Btu/hr.

QB = Burner input - Btu/hr.

QP = Pilot light (if used) input - Btu/hr.

*AFUE = Annual efficiency listed in the GAMA directory. (AFUE of electric furnaces installed indoors can be assumed to equal 1.00.)

Example: A house has a design heating load of 65,000 Btu/hr and is located in a city that experiences 3000 heating load hours. The furnace has an input rating of 80,000 Btuh and AFUE rating of 0.83, requires 0.5 KW for the fan and is equipped with electronic ignition.

$$\text{Btu/yr} = \frac{0.77 \times 3000 \times 65000}{0.83} = 180.9 \text{ Million}$$

Cu. ft. gas = 176500 (1025 Btu per cu. ft.)
Gal. oil = 1311 (138,000 Btu per gal.)

$$\text{Kwh/yr} = \frac{0.77 \times 3000 \times 65000}{0.83 \times 80000} \times 0.50 = 1311$$

A4-6 HSPF Rating

The annual Heating Season Performance Factor (HSPF) for heat pumps is listed in the "Directory of Certified Unitary Air Conditioners and Heat Pumps" published by the Air Conditioning and Refrigeration Institute (ARI).

In the following equation, the HSPF number can be used to estimate the annual heating energy that is consumed by the heat pump, the indoor and outdoor fans, the crank case heater and the controls. Note that the HSPF number includes an allowance for on/off cycling. Also note that the HSPF number does not apply to set-back control strategies nor should it be used to estimate the energy consumed by multi-zone systems.

$$\text{Heat Pump Kwh/yr} = \frac{0.77 \times HLH \times QH}{1000 \times HSPF}$$

Kwh/yr = Annual energy input to the heat pump.

0.77 = A factor which is required to make the design heating load hours correlate with the effective heating load hours experienced under actual operating conditions.

HLH = Heating load hours from map provided in the ARI standard 240 or as calculated by the following formula:

$$HLH = \frac{\text{Heating Degree Days} \times 24}{(65 - \text{Winter Design Temperature})}$$

QH = The design heat load - Btu/hr.

HSPF = Heating season efficiency.

Example: A house has a design heating load of 65,000 Btu/hr, and is located in a city that experiences 3000 heating load hours and the heat pump has an HSPF rating of 7.23.

$$\text{Kwh/yr} = \frac{0.77 \times 3000 \times 65000}{1000 \times 7.23} = 20,768$$

Input Btu/yr = 70,881,000 (3413 Btu per Kwh)

A4-7 SEER Rating

The Seasonal Efficiency Ratio (SEER) for air conditioners and heat pumps is listed in the "Directory of Certified Unitary Air Conditioners and Heat Pumps" published by ARI.

In the following equation, the SEER number can be used to estimate the annual cooling energy that is consumed by the cooling equipment, the indoor and outdoor fans, the crank case heater and the controls. Note that the SEER number includes an allowance for on/off cycling. Also note that the SEER number does not apply to set-up control strategies nor should it be used to estimate the energy consumed by multi-zoned systems.

$$\text{Cooling Kwh/Yr} = \frac{CLH \times QC}{1000 \times SEER}$$

Kwh/yr = Annual Cooling Energy Input.

CLH = Cooling load hours from map provided in the ARI standard 240 or as calculated by the following formula:

$$CLH = \frac{\text{Cooling Degree Days} \times 24}{(\text{Summer Design Temperature} - 65)}$$

QC = Design cooling (total) load - Btu/hr.

SEER = Cooling season efficiency

Example: A house has a design cooling load of 40,000 Btu/hr, and is located in a city that experiences 650 cooling load hours and the cooling equipment has a SEER rating of 8.00.

$$\text{Kwh/yr} = \frac{650 \times 40,000}{1000 \times 8.00} = 3,250$$

Input Btu/yr = 11,092,000   (3413 Btu per Kwh)

A4-8 Bin Calculations

The accuracy of bin calculations depend on the amount of detail (on weather patterns, envelope performance, occupancy and internal load schedules and equipment performance) included in the bin model. The example bin calculations in this section are only intended to illustrate the basic bin calculation procedure. However, the calculations shown here can be performed with a hand calculator in a reasonable amount of time and they do produce a rough estimate of annual energy requirements. Refer to Section A4-11 for a complete discussion of the parameters which should be considered in a comprehensive bin calculation.

A4-9 Bin Method - Furnace Energy

Since the envelope heating load varies directly with outdoor air temperature, the envelope heating energy (Btu/yr) associated with a given outdoor temperature bin equals the heating load (Btu/hr) at the bin midpoint temperature multiplied by the number of hours per year that the bin temperature is expected to occur. The annual envelope heating energy is equal to the sum of the energy that is associated with each temperature bin.

Figure A4-1 illustrates the procedure. To use this method, it is necessary to have weather data which tabulates the hours per year the various temperatures occur. This data can be obtained from the Government Printing Office. Ask for Engineering Weather Data AFM 88-29. For convenience, bin data for limited number of locations is provided in Table A4-1.

Figure A4-1 indicates that approximately 146 million Btu per year are required to heat the envelope. The fuel which must be purchased to provide this heat can be determined by using the following equation.

$$F = \frac{\text{Btu/yr.}}{E \times P}$$

Where:

Btu/yr. is the heat requirement as shown by Figure A4-1.

E is the seasonal efficiency of the equipment.

P is the heating value of the fuel.

F is the annual fuel consumption.

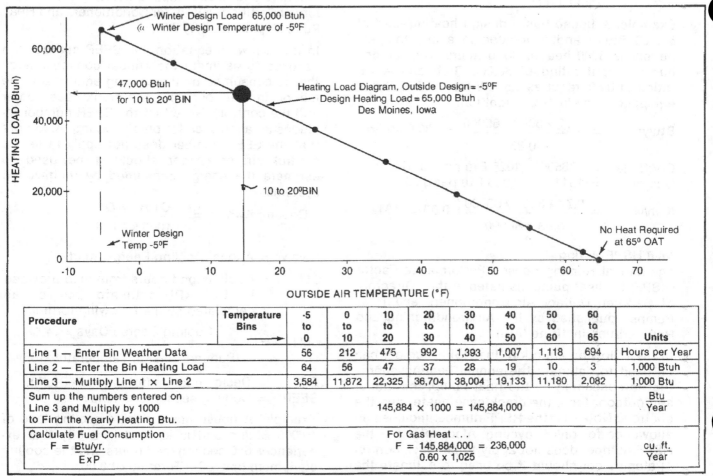

Procedure	Temperature Bins →	-5 to 0	0 to 10	10 to 20	20 to 30	30 to 40	40 to 50	50 to 60	60 to 65	Units
Line 1 — Enter Bin Weather Data		56	212	475	992	1,393	1,007	1,118	694	Hours per Year
Line 2 — Enter the Bin Heating Load		64	56	47	37	28	19	10	3	1,000 Btuh
Line 3 — Multiply Line 1 × Line 2		3,584	11,872	22,325	36,704	38,004	19,133	11,180	2,082	1,000 Btu
Sum up the numbers entered on Line 3 and Multiply by 1000 to Find the Yearly Heating Btu.		\multicolumn{8}{c}{145,884 × 1000 = 145,884,000}								Btu / Year
Calculate Fuel Consumption F = Btu/yr. / ExP		\multicolumn{8}{c}{For Gas Heat ... F = 145,884,000 / (0.60 × 1,025) = 236,000}								cu. ft. / Year

Figure A4-1 Bin Method for Furnace Energy

A4-10 Bin Method Heat Pump Energy

Heat pumps should be evaluated by the bin method because this method will account for changes in the coefficient of performance at various outside air temperatures and for the energy required by the auxiliary resistance heat. To use the bin method, the envelope load diagram must be constructed. Plotting the integrated capacity of the heat pump at various outside air temperatures is also necessary. Figure A4-2 illustrates the procedure. (Heat pump capacity at various outside air temperatures can be found in manufacturer catalogs.)

Heat Pump Energy Calculation Procedure

Line 1 Enter the hours per year for each bin.

Line 2 Enter the building load for each bin.

Line 3 Enter the integrated heat pump capacity for each bin.

Line 4 Subtract line 3 (heat pump output) from line 2 (building load) to find the load which must be handled by the auxiliary heating coils.

Line 5 Enter the KW input required by the heat pump for each bin. The KW requirements for various outside air temperatures are in the manufacturers data.

Line 6 Enter the heat pump running time for each bin. This is calculated by dividing line 2 by line 3. Use 1.0 for bins that have building loads which are greater than the heat pump capacity.

Line 7 Enter the heat pump Kwh for each bin. This is calculated by multiplying lines 1,5, and 6 together. Add the bin Kwh to find the yearly heat pump energy requirement.

Line 8 Enter the resistance heat KW for each bin. This is calculated by dividing line 4 by 3.413 (Btuh/Watt).

Line 9 Enter the resistance heat Kwh for each bin. This is calculated by multiplying line 1 by line 8. Total the bin Kwh to find yearly resistance coil energy requirement.

The total Kwh required is the sum of the heat pump Kwh and the resistance Kwh. In this example, the heat pump will require 13,869 Kwh/yr. and the auxiliary heaters will use 9,504 Kwh/yr., for a total of 23,373 Kwh/yr.

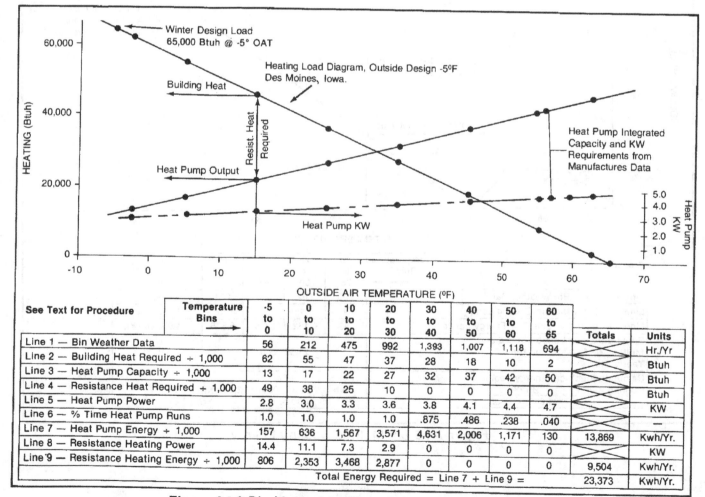

See Text for Procedure	Temperature Bins →	-5 to 0	0 to 10	10 to 20	20 to 30	30 to 40	40 to 50	50 to 60	60 to 65	Totals	Units
Line 1 — Bin Weather Data		56	212	475	992	1,393	1,007	1,118	694		Hr./Yr
Line 2 — Building Heat Required ÷ 1,000		62	55	47	37	28	18	10	2		Btuh
Line 3 — Heat Pump Capacity ÷ 1,000		13	17	22	27	32	37	42	50		Btuh
Line 4 — Resistance Heat Required ÷ 1,000		49	38	25	10	0	0	0	0		Btuh
Line 5 — Heat Pump Power		2.8	3.0	3.3	3.6	3.8	4.1	4.4	4.7		KW
Line 6 — % Time Heat Pump Runs		1.0	1.0	1.0	1.0	.875	.486	.238	.040		—
Line 7 — Heat Pump Energy ÷ 1,000		157	636	1,567	3,571	4,631	2,006	1,171	130	13,869	Kwh/Yr.
Line 8 — Resistance Heating Power		14.4	11.1	7.3	2.9	0	0	0	0		KW
Line 9 — Resistance Heating Energy ÷ 1,000		806	2,353	3,468	2,877	0	0	0	0	9,504	Kwh/Yr.
Total Energy Required = Line 7 + Line 9 =										23,373	Kwh/Yr.

Figure A4-2 Bin Method for Heat Pump Heating Energy.

A4-11 Bin Method Cooling Energy Estimate

The bin method can be used to estimate seasonal cooling energy requirement. To simplify the procedure, the seasonal energy efficiency rating (SEER) is used to estimate power consumption of compressor and auxiliaries. The design cooling load is taken directly from load calculation (Form J-1) and the cooling hours per year for bins are determined from weather data. Figure A4-3 illustrates the procedure for equipment that is operated in the cooling mode during the entire cooling season.

Air Conditioning Energy Calculation Procedure

Line 1 Enter the hrs./yr. for each bin.

Line 2 Enter the building load for each bin.

Line 3 Multiply Line 1 by Line 2 to find the envelope cooling energy per year for each bin. Add the energy required for each bin to find the total cooling energy required for the year (21,397,000 Btu/yr.)

Calculate the approximate equipment cooling energy.

$$\text{Equipment Kwh/yr.} = \frac{\text{(Btu/yr. from Line 3)}}{1000 \times \text{SEER}}$$

$$= \frac{21,397,000}{1,000 \times 9.5}$$

$$= 2,252 \text{ Kwh/yr. for cooling.}$$

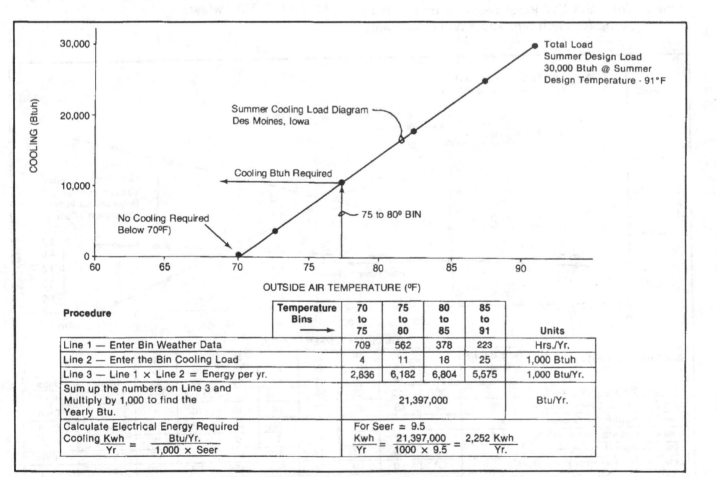

Procedure	Temperature Bins →	70 to 75	75 to 80	80 to 85	85 to 91	Units
Line 1 — Enter Bin Weather Data		709	562	378	223	Hrs./Yr.
Line 2 — Enter the Bin Cooling Load		4	11	18	25	1,000 Btuh
Line 3 — Line 1 × Line 2 = Energy per yr.		2,836	6,182	6,804	5,575	1,000 Btu/Yr.
Sum up the numbers on Line 3 and Multiply by 1,000 to find the Yearly Btu.			21,397,000			Btu/Yr.
Calculate Electrical Energy Required Cooling $\frac{\text{Kwh}}{\text{Yr}} = \frac{\text{Btu/Yr.}}{1,000 \times \text{Seer}}$		For Seer = 9.5 $\frac{\text{Kwh}}{\text{Yr}} = \frac{21,397,000}{1000 \times 9.5} = \frac{2,252 \text{ Kwh}}{\text{Yr.}}$				

Figure A4-3 Bin Method Cooling Energy

A4-12 Computer Analysis

The accuracy of any piece of computer software will depend on the detail included in the computerized model and the knowledge and skill of the individuals who create the program. Refer to the section below for a list of items that should be considered when Bin calculations or hourly calculations are computerized.

A4-13 Factors That Effect Energy Calculations

Residential energy consumption is effected by local weather patterns, the construction details of the building envelope, the operating characteristics of the heating and cooling system, the type of equipment and controls and the habits of the occupants. A comprehensive energy model should include provisions to estimate the impact of the following factors:

Weather
- Bin hours and corresponding dry bulb and wet bulb temperature
- Hourly and seasonal solar patterns
- Seasonal cloud cover
- Summer and winter wind velocity
- Correlation of the Bin temperatures with the solar, cloud and wind velocity patterns

Building Envelope
- Transmission loads
- Infiltration loads
- Solar loads
- Internal loads
- Ventilation loads (in some cases)
- Latent loads
- Thermal storage of construction materials
- Inter-zonal heat transfer
- Correlation of all loads with the weather patterns, occupancy patterns; equipment performance, HVAC system and control characteristics

Equipment Performance
- Output capacity at full load and at all part load operation operating conditions
- Fuel and/or power input requirements at full load and at all part load operating conditions
- On/off cycling penalties
- Equipment over-sizing penalties

- Energy required for intermittent or continuous operation of indoor fan
- Effect of latent loads on cooling equipment operating time
- Thermal balance points
- Economic balance points
- Auxiliary heat requirements during normal operation
- Auxiliary heat requirement during the defrost cycle
- Defrost cycle energy requirement
- Energy consumed by the outdoor fan(s)
- Energy which is not consumed by outdoor fans during defrost cycles
- Crank-case heater energy requirement
- Energy consumed by standing pilot lights
- Energy consumed by water pumps
- Duct losses
- Refrigerant line losses
- Benefits of variable and multi-speed motors
- Correlation of equipment performance with envelope loads and system controls

Controls
- Set points at design conditions
- Set up and set back set points
- Auxiliary and emergency heat controls
- Outdoor thermostat set point(s)
- On/off operating schedules
- Set up and set back schedules
- Equipment operating during recovery from set up or set back
- Defrost cycle control strategies
- Crank case heater control
- Fuel ignition controls
- Multi-zone controls
- Correlation with equipment performance, envelope loads, weather patterns and occupancy schedules.

Table A4-1
Bin Weather Data

AVERAGE NUMBER OF HOURS EACH TEMPERATURE SHOWN OCCURS IN A YEAR

OUTDOOR TEMPERATURE (F)	110 to 115	105 to 110	100 to 105	95 to 100	90 to 95	85 to 90	80 to 85	75 to 80	70 to 75	65 to 70	60 to 65	55 to 60	50 to 55	45 to 50	40 to 45	35 to 40	30 to 35	25 to 30	20 to 25	15 to 20	10 to 15	5 to 10	0 to 5	-5 to 0	-10 to -5	-15 to -10	-20 to -15	-25 to -20	-30 to -25	-35 to -30
Akron, Ohio			*	6	30	112	261	454	639	801	762	736	684	605	680	706	819	654	422	214	127	44	17	10	4	*				
Albany, N.Y.				10	31	115	230	387	559	741	769	720	694	694	677	751	814	552	376	254	174	90	62	20	2	2				
Albuquerque, N.M.		3	3	66	364	506	780	629	838	678	783	712	710	714	666	523	319	142	76	19	2					1				
Allentown, Pa.			*	12	76	153	305	471	657	812	783	792	741	710	712	848	523	501	248	129	61	23	7	3	1					
Amarillo, Tex.	1		24	129	288	381	475	607	804	824	752	671	714	684	612	595	499	315	191	115	52	34	5	1	2	1				
Anchorage, Alaska								8	50	180	499	930	995	741	637	713	831	795	662	560	376	285	211	124	94	46	28	5		
Atlanta, Ga.			2	23	172	361	620	880	1188	978	908	817	700	651	627	455	255	82	20	9	6	2								
Augusta, Ga.		1	17	88	274	531	652	959	1167	957	875	785	666	570	497	368	210	94	35	9	2	*								
Austin, Tex.		1	46	215	438	600	819	1251	1105	1032	768	717	567	472	355	220	78	51	13	15	2	*								
Bakersfield, California	*	18	100	254	371	474	613	742	631	898	966	977	908	746	541	247	77	7												
Baltimore, Md.			2	23	109	263	438	655	883	794	729	696	683	673	770	755	642	328	184	89	43	7	2							
Baton Rouge, La.			2	50	376	622	826	1307	1346	992	664	664	598	326	482	199	103	41	1	1										
Billings, Mont.		3	3	17	76	144	242	340	447	564	686	749	741	721	781	785	730	522	372	236	163	121	89	81	71	51	30	1		
Binghamton, New York				2	2	31	145	318	548	781	808	694	673	595	598	666	873	568	515	350	263	156	67	23	5	1				
Birmingham, Ala.			8	57	256	471	631	915	1166	1044	875	767	621	581	489	423	273	111	48	13	4	1								
Bismarck, N.D.			*	15	68	214	233	346	560	560	632	612	549	501	482	575	702	558	436	358	349	304	278	214	161	102	71	41	21	7
Boise, Idaho		*	6	39	135	307	376	484	609	663	790	748	854	809	748	865	881	498	272	142	47	18	4	9	*					
Boston, Mass.			2	10	93	127	245	433	676	819	804	781	766	757	828	848	674	429	256	151	74	35	4							
Brownsville, Tex.			3	57	489	862	1579	1901	1419	963	564	390	270	187	73	25	13													
Buffalo, N.Y.				8	15	76	232	396	755	819	875	767	621	581	684	796	821	605	427	230	125	66	19	2	*	*				
Burbank, Calif.	7	41	129	257	431	565	808	1163	1562	1562	1186	752	727	661	292	83	10													
Burlington, Ia.		*	9	48	129	199	377	584	743	753	714	613	528	485	572	708	797	542	356	233	161	108	55	29	13	3				
Burlington, Vermont		3	31	95	189	362	573	670	703	655	694	603	637	716	752	561	491	332	272	101	60	10	17	39	2					
Calgary, Alta.			9	53	28	89	173	594	831	698	558	694	839	781	614	760	581	218	198	280	109	36	9	23	49	75	35	12		
Casper, Wyoming	3		3	66	201	283	347	273	532	592	606	642	670	1484	782	806	324	683	476	303	132	116	73	45	30	15	2	2		
Charleston, S.C.			1	34	148	436	736	1156	1311	1134	1006	792	833	546	433	673	755	626	448	262	136	75	43	18	7					
Charleston, W.V.				3	57	270	471	606	912	949	767	689	661	667	607	633	630	356	252	135	73	22	7	1						
Charlotte, N.C.			5	52	203	397	567	747	908	814	839	752	730	684	634	515	360	166	87	23	5	7	1	4						
Chattanooga, Tenn.			16	87	260	535	713	930	986	930	776	746	696	631	538	423	273	215	87	32	10	1								
Cheyenne, Wyo.				3	32	93	214	301	409	514	637	771	769	765	883	810	797	608	499	279	177	92	55	34	13	4	5			
Chicago, Ill.			4	32	96	209	355	531	697	767	676	581	617	561	616	826	842	517	337	207	117	79	55	18	11	3				
Cincinnati, Ohio				24	105	250	431	613	822	855	738	692	661	619	642	692	709	475	218	101	60	36	10	6	1					
Cleveland, Ohio			5	8	56	155	311	567	709	831	783	597	592	513	445	569	657	581	280	198	109	81	56	22	9					
Cold Bay, Alaska								2	6	31	211	1042	1484	725	1327	1755	1245	673	476	303	132	39	5							
Colorado Springs, Col.		2	1	46	176	276	397	477	600	784	805	760	725	678	673	755	626	448				75	43	18	7					
Columbia, S.C.			17	93	279	502	655	964	1126	941	838	778	661	564	546	394	286	115	36	12										
Columbus, Ohio			5	18	83	270	392	567	747	814	737	701	655	667	661	633	739	537	177	136	71	25		6	5	4				
Corpus Christi, Texas				16	36	260	535	747	986	1041	921	551	444	302	180	83	27	34	87	3	1									
Dallas, Tex.		9	79	273	493	659	880	942	831	795	693	629	712	576	504	371	231	91	34	17	4									
Dayton, Ohio			12	83	65	202	402	607	817	832	717	627	678	601	576	698	786	558	309	182	99	55	24	16	5	1	*			
Denver, Colorado		9	42	103	236	332	437	549	684	783	783	627	678	704	658	717	721	553	359	216	119	55	23	16	6	5				
Des Moines, Ia.			12	30	94	148	378	562	709	783	694	597	637	513	492	630	657	583	439	280	195	131	81	56	26	9				
Detroit, Michigan			5	30	47	148	314	516	721	783	695	633	592	566	595	808	884	618	377	248	131	61	17	4						
Duluth, Minn.			1		4	23	86	194	323	458	597	574	613	605	528	585	766	638	617	499	373	284	229	190	131	142	50	20		
Edmondton, Alta.		2		2	4	23	78	182	403	750	921	1006	814	680	515	644	863	571	426	404	338	281	196	154	105	64	19	10	4	
El Paso, Tex.	2		42	204	406	586	740	865	970	893	759	759	712	680	514	342	205	90	35	8	3	1								
Evansville, Ind.			12	83	199	346	530	721	811	735	670	658	678	643	658	683	654	380	146	81	47	20	9	4		*				
Fairbanks, Alaska						14	54	118	202	361	515	658	637	513	429	409	495	455	457	447	401	379	401	237	332	270	206	159	126	121
Fargo, N.D.		*		6	30	108	216	362	505	626	680	641	574	449	565	569	657	578	439	360	255	274	75	124	182	124	70	33	19	2
Flint, Michigan				1	30	108	216	253	411	745	695	641	574	449	565	675				248	208	88	34	17	4		*			
Fort Wayne, Ind.			1	15	80	180	353	502	667	762	682	664	613	586	639	754	890	615	366	183	101	61	32	15	5	2				
Fort Worth, Texas		5	54	246	440	596	788	982	889	774	689	622	648	591	538	422	294	132	44	12	3	1								
Fresno, California			56	192	297	392	490	607	709	803	921	1006	1036	952	673	426	168	34	5											
Galveston, Tex.					38	56	280	415	594	718	729	671	589	564	589	794	983	701	454	256	157	67	28	6	2		1			
Grand Rapids, Mich.		2		15	80	122	180	353	502	762	682	664	613	586	639	754	890	615	366	183	101	61	32	15	5	2				

| OUTDOOR TEMPERATURE (F) | 110 to 115 | 105 to 110 | 100 to 105 | 95 to 100 | 90 to 95 | 85 to 90 | 80 to 85 | 75 to 80 | 70 to 75 | 65 to 70 | 60 to 65 | 55 to 60 | 50 to 55 | 45 to 50 | 40 to 45 | 35 to 40 | 30 to 35 | 25 to 30 | 20 to 25 | 15 to 20 | 10 to 15 | 5 to 10 | 0 to 5 | -5 to 0 | -10 to -5 | -15 to -10 | -20 to -15 | -25 to -20 | -30 to -25 | -35 to -30 |

*Less than one hour

Source of data: Lennox Ind.

Table A4-1 (Continued)

AVERAGE NUMBER OF HOURS EACH TEMPERATURE SHOWN OCCURS IN A YEAR

OUTDOOR TEMPERATURE (F)	110 to 115	105 to 110	100 to 105	95 to 100	90 to 95	85 to 90	80 to 85	75 to 80	70 to 75	65 to 70	60 to 65	55 to 60	50 to 55	45 to 50	40 to 45	35 to 40	30 to 35	25 to 30	20 to 25	15 to 20	10 to 15	5 to 10	0 to 5	-5 to 0	-10 to -5	-15 to -10	-20 to -15	-25 to -20	-30 to -25	-35 to -30
Great Falls, Mont.					36	86	160	270	374	501	635	805	884	847	822	809	712	553	383	231	154	112	92	97	58	49	55	15	5	1
Green Bay, Wisconsin				5	9	34	135	331	474	522	758	598	720	522	542	649	820	689	515	373	321	231	160	95	42	19	3			
Greensboro, N.C.				2	33	135	318	504	659	963	846	817	721	686	621	614	466	261	92	36	15	12								
Halifax, N.S.						2	29	105	251	547	956	941	884	809	799	998	959	528	353	245	174	81	44	12	3					
Harrisburg, Pa.				14	61	189	345	545	790	823	753	757	685	682	758	901	766	552	370	233	153	77	33	11	3					
Hartford, Connecticut				3	29	118	274	419	617	755	751	690	649	575	683	807	825	645	389	286	211	153	86	99	73	69	61	28	10	
Helena, Mont.					5	24	92	164	234	344	465	560	690	761	776	826	900	809	645	389	286	211	180	107	79	73	61			
Hilo, Hawaii											11																			
Honolulu, Hawaii					1	24	975	1552	2126	3001	3458	2789	2272	722	361	122	6													
Houston, Tex.			1	53	323	676	922	1621	1263	722	796	684	553	396	220	100	50	25	9	4										
Huron, South Dakota					103	205	238	443	554	624	614	569	513	488	502	574	652	571	476	419	305	262	208	145	83	56	21	7		1
Indianapolis, Ind.			2	23	102	318	420	604	740	763	708	601	620	614	653	751	752	543	279	135	74	53	28	7	3					
Jackson, Miss.			16	143	383	546	719	1051	1196	989	810	669	605	557	485	325	176	64	17	4	3	1								
Jacksonville, Fla.			2	70	316	607	949	1658	1329	979	895	686	598	350	660	262	52	1												
Kansas City, Mo.	1	8	36	107	221	367	687	687	767	740	593	617	598	581	660	643	600	389	251	147	93	51	23	7						
King Salmon, Alaska					30	121	256	409	596	725	706	696	628	592	575	741	920	734	451	299	158	76	33	12	2					4
Knoxville, Tenn.			3		626	801	1161	1346	958	838	687	563	471	343	192	82	27													
LaCrosse, Wis.				32	156	335	578	793	1042	935	847	709	676	698	639	585	442	209	60	26	15	5		7		29	19	8		1
LaGuardia, N.Y.			2	14	32	103	225	458	638	754	738	655	520	481	518	454	442	209	60	26	15	5								
Lake Charles, La.			2	40	51	170	834	882	771	765	814	768	622	450	825	804	507	262	139	69	22	6	1							
Lansing, Mich.			3	30	121	640	834	1444	1302	1038	838	717	628	592	303	126	58	14	6	5			2	12						
Laredo, Texas		6	189	469	626	801	1161	1346	958	838	687	563	471	343	192	82	27													
Las Vegas, Nevada	16	101	301	431	474	602	678	669	651	644	699	786	769	716	591	396	194	44	7											
Lexington, Kentucky			11	100	124	263	274	630	898	898	710	656	644	611	629	627	654	441	238	144	80	35	16	7						
Little Rock, Ark.				288	125	471	691	944	996	847	782	704	652	702	603	469	306	136	41	23	2	1					1			
Los Angeles, Calif.		1		4		18	68	202	670	1458	2331	2130	1183	532	151	15						1	2	2		1				
Louisville, Ky.		6	64	55	32	350	538	717	831	757	702	700	642	663	620	703	631	324	137	66	37	8		2		2		1		
Lubbock, Texas		14	14	209	33	544	708	833	713	688	829	642	618	546	504	623	490	346	180	86	33	5	7	8	50	46		6		
Macon, Georgia	2	3	109	85	41	538	665	995	1239	905	800	755	648	545	487	362	210	109	39	11	3	5	2	13	19					
Madison, Wis.			5	42	131	264	455	586	680	724	636	565	528	539	896	690	659	307	215	171	60	93	7	31	13	25		1		
Medford, Ore.		6	55	124	207	274	329	443	566	715	901	1000	1086	1143	984	660	984	659	48	7	1	1	1							
Memphis, Tenn.			17	120	471	672	669	651	855	716	716	645	71	602	591	396	442	224	54	12	7									
Miami, Fla.		1		5	125	297	277	630	810	898	452	277	147	71	26	4														
Midland, Texas			28	177	888	888	1795	2463	1708	810	452	678	631	669	603	451	337	163	80	23	5	1					1			
Milwaukee, Wis.				6	32	532	673	865	914	793	720	678	634	585	838	774	913	459	421	285	176	116	83	47	18	4	2	8	3	
Minneapolis, Minn.				7	41	128	226	268	597	753	735	666	657	585	522	623	666	598	526	357	288	219	177	124	72	46	50		8	
Missoula, Mont.		14	7	6	159	97	223	390	713	418	553	591	666	504	468	952	1016	736	465	284	170	89	43	38	19	9	2			
Mobile, Ala.			2	38	523	523	785	1338	1466	1135	898	567	551	537	498	180	74	23	2	4		10								
Moline, Ill.		1	19	88	209	500	574	909	694	744	799	591	708	567	370	543	312	108	376	247	175	111	66	39	13	8				
Montgomery, Ala.			7	89	332	500	751	1235	1239	945	799	708	551	455	551	827	159	232	17	5	3	2								
Montreal, Que.					43	103	751	593	586	593	724	666	646	589	555	706	719	523	437	355	318	280	180	107	79	4		3		
Nashville, Tenn.				76	220	404	583	804	919	876	836	729	644	601	643	562	444	241	89	43	22	5	2		21					
Newark, New Jersey		2	13	21	79	200	383	615	814	755	697	719	556	448	784	143	18	7	4	1	38	11	2	54	7	6				
New Orleans, La.			2	16	259	606	966	1259	1047	910	799	719	544	556	844	637	603	330	188	109	38	89	88							
New York, New York				5	28	96	263	604	926	877	754	745	745	703	838	858			2		26	10	2							
Norfolk, Va.		1	20	113	273	481	909	1113	995	808	858	795	681	782	737	543	312	108	6	4		9		*						
Oklahoma City, Okla.		20	89	240	421	606	909	847	769	684	799	725	683	601	609	498	401	232	135	66	19	2	2	2						
Oakland, California			1	2	26	61	135	339	756	1498	2431	2431	1858	971	509	155	20	*												
Omaha, Neb.			14	45	134	263	435	579	703	743	637	544	527	567	584	672	683	521	374	261	180	127	88	54		21			3	
Orlando, Florida		1		5	666	225	1004	1732	1680	1123	830	564	377	245	156	30						9								
Philadelphia, Pa.		1	17	74	225	404	420	655	864	808	735	663	644	548	555	654	654	335	189	100	32		64	15	10		6			
Phoenix, Arizona		129	324	507	611	698	798	762	722	910	799	749	637	659	540	391	57	8				30	30	7	7					
Pittsburgh, Penn.				17	88	606	798	503	722	814	799	678	688	587	631	688	774	569	462	250	159	94	58	34	38	19	9	2		
Pocatello, Idaho					98	96	311	359	467	536	585	638	688	703	743	835	891	678	360	250	151	94	30	7	10	10		8		
Portland, Me.	15		1	20	14	61	129	263	387	642	808	857	773	703	805	841	271	384	573	271	190	117	64	15	9		4			1

| OUTDOOR TEMPERATURE (F) | 110 to 115 | 105 to 110 | 100 to 105 | 95 to 100 | 90 to 95 | 85 to 90 | 80 to 85 | 75 to 80 | 70 to 75 | 65 to 70 | 60 to 65 | 55 to 60 | 50 to 55 | 45 to 50 | 40 to 45 | 35 to 40 | 30 to 35 | 25 to 30 | 20 to 25 | 15 to 20 | 10 to 15 | 5 to 10 | 0 to 5 | -5 to 0 | -10 to -5 | -15 to -10 | -20 to -15 | -25 to -20 | -30 to -25 | -35 to -30 |

*Less than one hour

Source of data: Lennox Ind.

Table A4-1 (Continued)

OUTDOOR TEMPERATURE (F)	-35 to -30	-30 to -25	-25 to -20	-20 to -15	-15 to -10	-10 to -5	-5 to 0	0 to 5	5 to 10	10 to 15	15 to 20	20 to 25	25 to 30	30 to 35	35 to 40	40 to 45	45 to 50	50 to 55	55 to 60	60 to 65	65 to 70	70 to 75	75 to 80	80 to 85	85 to 90	90 to 95	95 to 100	100 to 105	105 to 110	110 to 115
Portland, Ore.							1	2	4	15	31	52	97	301	702	1291	1313	1312	1359	959	546	367	208	128	53	21	3	3		
Providence, R.I.						1	12	12	34	66	134	257	418	709	801	690	822	824	827	844	803	664	419	228	73	22	2			
Pueblo, Colo.				1		8	26	39	52	117	203	365	496	570	556	674	674	677	708	714	714	587	383	296	210	68	10			
Raleigh, N.C.										7	23	72	197	383	520	589	648	728	813	915	968	1049	726	518	356	174	49	5		
Rapid City, S.D.				5	10	46	79	128	194	246	273	420	594	742	694	675	621	632	666	655	590	488	370	275	198	108	52	10		
Regina Sask.	12	27	51	99	168	221	238	281	330	349	441	533	563	648	559	530	530	616	693	533	481	386	282	168	92	47	10			
Reno, Nevada							4	15	37	101	227	530	733	829	890	890	845	710	572	477	371	418	333	243	120	35	2			
Richmond, Va.									13	28	58	108	269	497	617	688	695	710	759	812	854	914	726	483	330	163	59	12		
Roanoke, Virginia							6	30	11	28	86	186	307	533	702	710	712	695	704	758	924	928	620	459	305	91	8			
Rochester, N.Y.							6	30	69	135	233	403	606	823	816	710	661	691	746	750	712	590	378	241	125	43	9			
Sacramento, Calif.													8	93	355	701	1049	1298	1329	1071	773	486	375	375	276	192	93	34	5	*
Salem, Oregon									*	3	20	46	172	435	707	1100	1049	1352	1163	833	540	405	261	169	100	47	14	5	*	
Salt Lake City, Utah								16	46	90	150	308	572	842	841	725	652	690	661	632	641	588	461	343	272	150	43	3		
San Antonio, Tex.										1	8	15	36	81	161	340	407	529	675	853	1019	1180	1310	863	623	427	215	22		
San Diego, Calif.													15	121	517		1319	972	1863	2389	1821	832	270	99	22	4				
San Francisco, Calif.															88	393	1319	2469	2278	1174	571	192	65	31	10	2				
San Juan, P.R.																				158	1450	3780	2521	837	19					
Savannah, Ga.											2	9	35	106	235	367	496	586	736	935	1054	1313	1279	820	488	206	66	8		
Scranton, Pennsylvania						2	9	31	78	178	264	392	575	848	805	628	621	629	719	784	666	413	254	83	17	4				
Seattle, Wash.									*	5	21	42	133	536	1121	1605	1337	1295	1132	699	406	233	105	36	17	4	1			
Shreveport, La.							1			3	8	25	43	161	306	464	584	658	695	811	924	1065	1112	767	583	394	133	18		
Sioux City, Ia.		2		14	38	73	97	152	247	361	468	595	732	642	550	523	527	575	642	701	618	507	365	207	97	26	4			
Sioux Falls, S.D.		16	2	43	59	102	152	208	293	448	520	585	870	625	501	526	522	605	649	684	566	443	277	155	67	7				
South Bend, Indiana				*	8	16	47	81	166	250	449	661	870	694	564	526	567	608	728	806	698	507	318	150	41	3				
Spokane, Wash.			*	3	6	29	52	91	153	302	625	1060	974	853	805	786	715	633	525	414	294	212	136	66	16	1				
Springfield, Ill.				3	20	34	68	88	150	273	531	832	688	623	588	569	573	651	757	725	645	486	257	152	39	7	1			
Springfield, Missouri					3	14	29	68	139	235	437	588	621	616	621	602	759	846	876	647	475	331	169	60	7	1				
St. John, N.B.			2	1	42	97	128	201	264	335	427	669	986	784	692	796	941	1022	887	390	207	79	11							
Syracuse, N.Y.			1	1	7	10	59	79	141	241	370	564	800	752	708	669	709	753	742	752	604	406	255	109	28	4				
Tallahassee, Florida									9	57	126	219	331	428	568	760	895	1061	1618	1275	739	530	149	4						
Tampa, Fla.												1	6	48	137	216	345	570	877	1187	1387	1910	1126	752	195	6				
Toledo, Ohio						7	12	28	50	113	214	337	597	920	811	640	636	611	606	677	782	491	306	169	61	18	*			
Topeka, Kan.					2	10	28	58	105	173	280	500	681	654	601	581	590	707	625	729	723	649	490	334	214	110	34	7		
Toronto, Ont				2	14	45	98	195	291	339	413	600	872	745	616	577	581	700	691	701	577	387	224	84	14					
Tucson, Arizona												25	98	248	417	598	716	763	781	781	870	959	777	656	520	357	152	31	*	
Tulsa, Oklahoma							1	2		29	75	159	265	386	535	611	637	622	636	671	752	816	816	649	481	333	149	43	10	
Vancouver, B.C.									12	5	13	21	110	386	753	1364	1618	1321	1217	936	559	310	108	18	3					
Waco, Texas										1	3	24	84	216	354	501	558	651	622	701	830	909	1101	829	612	482	249	42	1	
Wake Island, Pacific																					5	621	3336	3944	863					
Washington D.C.										11	48	104	213	532	768	798	715	690	723	823	835	947	735	496	284	101	31	1		
West Palm Beach, Fla.														2		57	115	202	291	455	677	1672	2413	1664	860	183	3			
Wichita Falls, Texas								4		10	27	114	228	388	497	581	636	627	606	677	714	784	825	741	560	406	260	83	2	
Wichita, Kan.					3		22	47	78	140	236	349	549	614	618	658	669	670	757	804	768	673	503	341	213	105	47	11	
Wilmington, Del.							6	7	39	107	200	369	667	816	752	476	648	731	721	702	630	390	209	65	17	2				
Winnipeg, Man.	8	17	60	134	187	248	278	303	333	361	406	454	505	667	532	451	476	552	637	702	703	290	152	65	17	2				
Winston Salem, N.C.							1	1	4	18	43	116	245	456	622	708	704	710	694	801	907	1086	703	518	323	105	7			
Yakima, Wash.				1	1	6	1	31	63	118	242	304	629	904	807	841	876	823	752	694	582	450	342	271	165	86	31	2		
Youngstown, Ohio					3	3	10	31	63	158	242	420	645	840	681	645	620	664	698	773	821	641	421	266	102	21	2			

| OUTDOOR TEMPERATURE (F) | -35 to -30 | -30 to -25 | -25 to -20 | -20 to -15 | -15 to -10 | -10 to -5 | -5 to 0 | 0 to 5 | 5 to 10 | 10 to 15 | 15 to 20 | 20 to 25 | 25 to 30 | 30 to 35 | 35 to 40 | 40 to 45 | 45 to 50 | 50 to 55 | 55 to 60 | 60 to 65 | 65 to 70 | 70 to 75 | 75 to 80 | 80 to 85 | 85 to 90 | 90 to 95 | 95 to 100 | 100 to 105 | 105 to 110 | 110 to 115 |

*Less than one hour

AVERAGE NUMBER OF HOURS EACH TEMPERATURE SHOWN OCCURS IN A YEAR

Appendix A-5
Detailed Infiltration Estimate

The Table 5 infiltration estimates are based on the procedure outlined below subject to the following assumptions:
- Length to width ratio of house between 1:1 and 3:1.
- Total glass and door area between 10 and 30 percent of the wall area.
- All walls completely exposed to the wind
- One kitchen and two bathroom exhaust fans; one dryer vent.
- Fossil furnace and water heater.
- Ducts in unconditioned space.
- Two recessed lighting fixtures, two pipe penetrations, one duct penetration, no conditioning equipment installed in a window or wall.
- Leakage areas per square foot of exposed area:

	Best	Average	Poor
Walls	.008	.032	.054
Windows and Glass Doors	.018	.040	.070
Panel and French Doors	.050	.100	.180

When the conditions that are outlined above do not apply, the infiltration calculation procedure (on the following page) can be used to generate air change estimates that can be used in place of the Table 5 values.

The procedure presented here is based on the information published in the 1985 ASHRAE Fundamentals (Section 22 pages 22.13 through 22.17). Also see Figure 8 and 9 (pages 22.7 and 22.8 and page 22.11). Other information incorporated in this table was extracted from:

Lbl - 16221; Lawrence Berkeley Publication "Component Leakage Areas in Residential Buildings". (Reinhold and Sonderegger)

Reports on Infiltration Barriers, Submitted to U.S. Department of Housing and to the Florida Energy Commission (Steven Winter and Associates, NY, NY).

The Florida and California Energy Codes.

Industrial Report on the Effects of Sheathing Products on Wall Infiltration; published by Simplex Industries (DeNunzio and Strass).

Information on Air Infiltration Barriers Provided by Monsanto, St. Louis, Missouri.

DETAILED INFILTRATION ESTIMATE

Step 1: Calculate the Leakage Area (sq. in.) for the Entire House.

	Leakage Area Factors (af)			"A" Enter (af) Value	"B" Enter Area or Count	"C" Multiply (af) x Area
	Best	Average	Poor			
Exterior Walls						
Wood Framing & Exterior Siding					Gross	C = A x B
Slats, Boards	.007	.038	.068	/////	Wall	/////
Panels	.007	.025	.044	/////	Area	/////
Masonry Walls, Any Siding	.009	.032	.054	_____	_____	_____
Windows & Sliding Glass Doors					Windows	
With Storm	.015	.030	.040	/////	Area	/////
No Storm	.020	.050	.080	_____	_____	_____
Jalousie Windows with Storm	NA	NA	.080	/////	Area	/////
Jalousie Windows no Storm	NA	NA	.150	_____	_____	_____
Wood Metal and French Doors					Door	
With Storm	.040	.080	.120	/////	Area	/////
No Storm	.060	.120	.240	_____	_____	_____
Envelope Penetrations					Count	
Recessed Light Fixtures	1.00	NA	2.00	_____	_____	_____
Pipe Penetrations	.16	NA	.93	_____	_____	_____
Duct Penetrations	.25	NA	3.70	_____	_____	_____
Window or Thru Wall Equipment	NA	2.00	4.00	_____	_____	_____
Vents and Exhausts					Count	
Kitchen Fan	.80	NA	6.00	_____	_____	_____
Bathroom Fan	1.70	NA	3.10	_____	_____	_____
Dryer Vent	.47	NA	4.00	_____	_____	_____
Furnace, Ducts and Water Heater					Count	
Furnace	0	5.00	10.00	_____	_____	_____
Water Heater	NA	NA	3	_____	_____	_____
Ducts in Uncond Space, No Tape	NA	NA	22	_____	_____	_____
					Count	
Fireplace	6.00	12.00	54.00	_____	_____	_____

Leakage Area Equals the Sum of Column "C" Values = _____ sq. in.

Step 2: Select the Winter & Summer CFM/sq. in. Factors.

	Winter CFM/sq. in. Factor			Summer CFM/sq. in. Factor		
	Number of Stories			Number of Stories		
Wind Exposure	One	Two	Three	One	Two	Three
Less Than 30% Exposed Wall Area	1.3	1.7	2.0	.4	.5	.6
30% to 60% Exposed Wall Area	1.6	2.0	2.3	.6	.7	.8
More Than 60% Exposed Wall Area	1.9	2.3	2.6	.8	.9	1.0

Winter CFM/sq. in. = _____ Summer CFM/sq. in. = _____

Step 3: Calculate the Infiltrations AC/HR for Winter & Summer

	Leakage Area: sq. in.	Infilt' CFM Per sq. in.	Min. Per Hr.	Volume * Cu. Ft.	Crack AC/Hr	Door Traffic AC/Hr	Total AC/Hr	
Winter =	_____ x	_____ x	60 /	_____ =	_____	+ .10	= _____	Heating
Summer =	_____ x	_____ x	60 /	_____ =	_____	+ .10	= _____	Cooling

*Volume Equals Total Floor Area of Conditioned Space Multiplied by the Average Ceiling Height.

Table A5-1

Building Component Evaluation

Walls

Best: Continuous infiltration barrier installed on outside of the walls which covers the entire wall area including the sills, joists, window and door frames and corners and which has all seams, cuts and penetrations lapped and taped. (Note that when an infiltration barrier is installed, the presence of a vapor barrier is redundant as far as infiltration is concerned. However, a suitable vapor barrier will still be required to prevent condensation and ice formation within the wall.)

Average: No infiltration barrier, some cuts and penetrations in plastic film vapor barrier. Some cracks at sills, joists, window and door frames and corners not completely covered with plastic vapor barrier and (or) sealed with mastic, joints at sheathing panels nailed tight but not sealed with mastic.

Poor: No infiltration barrier or plastic vapor barrier, insulation blankets that have impermeable backing not lapped and taped, no attempt to seal cracks between framing components or between sheathing panel joints.

Windows, Glass Doors and Doors

Best: Fixed glass, movable glass or doors that are certified to have a tested leakage of less than 0.25 CFM per running foot of crack.

Average: Movable glass or doors that are weather stripped and appear to provide tight closure or which are certified to have a tested leakage of between 0.25 and 0.50 CFM per running foot of crack.

Poor: Movable glass or doors that are not weather stripped and (or) which that do not appear to provide tight closure or which are not certified to have a tested leakage of 0.50 CFM per running foot of crack or less.

Penetrations

Best: Cracks around envelope penetrations for plumbing, gas service, telephone, electrical service and HVAC equipment, sealed with flexible mastic that will not degrade with time or changes in temperature. Recessed light fixtures sealed by a gasket.

Poor: No attempt to seal cracks, or cracks around penetrations sealed with mastic that pulls away from surfaces or which hardens and breaks away in pieces after being exposed to the weather. No attempt to seal recessed light fixtures.

Fans and Vents

Best: Vents and exhaust fans equipped with tight fitting backdraft dampers.

Poor: Vents and exhaust fans not equipped with tight fitting backdraft dampers.

Fireplaces

Best: Combustion air from outdoors, air intake and flue equipped with tight damper and glass doors installed in front of fireplace.

Average: Combustion air from the conditioned space, flue equipped with tight damper and (or) tight fitting glass doors installed in front of fireplace.

Poor: Combustion air from the conditioned space, no flue damper (or doors) or poorly fitted flue damper.

Furnaces

Best: Combustion air from the outdoors.

Average: Combustion air from the indoors, flue equipped with stack damper.

Poor: Combustion air from the indoors, flue not equipped with stack damper.

Wind Exposure

"Exposed wall" refers to walls that are exposed to the direct force of the wind. No structures, trees or obstructions within 30 feet of the wall.

Walls that are not exposed must be shielded from the wind by structures, trees, shrubs, or other obstructions that are located within 30 feet of the house. Do not take credit for plants or structures which can be easily removed or relocated.

Air Infiltration Barrier

An air infiltration barrier is defined as a continuous layer of water proof material placed on the outside of the exterior wall framing to restrict the inward air leakage while permitting the outward escape of vapor transmission from the wall cavity. The barrier must be lapped and taped so as to cover the entire wall area including all sills, joists, corners, window and door frames etc. and must not be degraded by cuts or punctures.

EXAMPLE INFILTRATION CALCULATION

Step 1: Calculate the Leakage Area (sq. in.) for the Entire House.

	Leakage Area Factors (af) Best	Average	Poor	"A" Enter (af) Value	"B" Enter Area or Count	"C" Multiply (af) x Area
Exterior Walls						
Wood Framing & Exterior Siding					Gross	C = A x B
Slats, Boards	.007	.038	.068	/////	Wall	/////
Panels	.007	.025	.044	/////	Area	/////
Masonry Walls, Any Siding	.009	.032	.054	.068	1440	97.92
Windows & Sliding Glass Doors					Windows	
With Storm	.015	.030	.040	/////	Area	/////
No Storm	.020	.050	.080	.080	216	17.28
Jelosie Windows with Storm	NA	NA	.080	/////	Area	/////
Jelosie Windows no Storm	NA	NA	.150	NA	NA	NA
Wood Metal and French Doors					Door	
With Storm	.040	.080	.120	/////	Area	/////
No Storm	.060	.120	.240	.240	40	9.60
Envelope Penetrations					Count	
Recessed Light Fixtures	1.00	NA	2.00	2.00	4	8.00
Pipe Penetrations	.16	NA	.93	.93	1	.93
Duct Penetrations	.25	NA	3.70			
Window or Thru Wall Equipment	NA	2.00	4.00			
Vents and Exhausts					Count	
Kitchen Fan	.80	NA	6.00	6.00	1	6.00
Bathroom Fan	1.70	NA	3.10	3.10	2	6.20
Dryer Vent	.47	NA	4.00	4.00	1	4.00
Furnace, Ducts and Water Heater					Count	
Furnace	0	5.00	10.00	10.00	1	10.00
Water Heater	NA	NA	3	3.00	1	3.00
Ducts in Uncond Space, No Tape	NA	NA	22	22.00	1	22.00
					Count	
Fireplace	6.00	12.00	54.00	54.00	1	54.00

Leakage Area Equals the Sum of Column "C" Values = __238.9__ sq. in.

Step 2: Select the Winter & Summer CFM/sq. in. Factors.

	Winter CFM/sq. in. Factor			Summer CFM/sq. in. Factor		
	Number of Stories			Number of Stories		
Wind Exposure	One	Two	Three	One	Two	Three
Less Than 30% Exposed Wall Area	1.3	1.7	2.0	.4	.5	.6
30% to 60% Exposed Wall Area	1.6	2.0	2.3	.6	.7	.8
More Than 60% Exposed Wall Area	1.9	2.3	2.6	.8	.9	1.0

Winter CFM/sq. in. = __2.0__   Summer CFM/sq. in. = __0.70__

Step 3: Calculate the Infiltration AC/HR for Winter & Summer.

	Leakage Area: sq. in.	Infilt' CFM Per sq. in.	Min. Per Volume* Hr. Cu. Ft.	Crack AC/Hr	Door Traffic AC/Hr	Total AC/Hr
Winter =	238.9	x 2.0	x 60 / 14,400 =	1.99	+ .10	= 2.09 Heating
Summer =	238.9	x 0.70	x 60 / 14,400 =	0.70	+ .10	= 0.80 Cooling

*Volume Equals Total Floor Area of Conditioned Space Multiplied by the Average Ceiling Height.
900 x 8 + 900 x 8 = 14,400 cu. ft.

Figure A5-1

List of Abbreviations

ACCA	-	Air Conditioning Contractors of America
NAHB	-	National Association of Home Builders
ASHRAE	-	American Society of Heating, Refrigerating, and Air Conditioning Engineers
ARI	-	Air-Conditioning & Refrigeration Institute
NFPA	-	National Fire Protection Association
FHA	-	Federal Housing Authority
HUD	-	Department of Housing and Urban Development
AFM	-	Air Force Manual
HVAC	-	Heating, Ventilating, and Air Conditioning
GAMA	-	Gas Appliance Manufacturers Association

List of Symbols

AC	-	Air Change
AC/HR	-	Air Change Per Hour
AFC	-	Annual Fuel Consumption
AFUE	-	Annual Fuel Efficiency Rating
Btu	-	British Thermal Unit
Btuh	-	Btu Per Hour
CFM	-	Cubic Feet Per Minute
CLH	-	Cooling Load Hours
cu. ft.	-	Cubic Foot
D.D.	-	Degree Days
gr diff	-	Difference in Grains of Humidity
E	-	Efficiency of Equipment
e	-	Emittance
ETD	-	Equivalent Temperature Difference - °F
°F	-	Degree Fahrenheit
GAL	-	Gallon
H	-	Convective Heat Transfer Coefficient
HL	-	Heating Load
HLH	-	Heating Load Hours
hr	-	Hour
HSPF	-	Heating Season Performance Factor
HTM	-	Heat Transfer Multiplier
Kwh	-	Kilowatt Hour
L	-	Length - Inches
lb.	-	Pound
LF	-	Linear Foot
MZ	-	Multi-Zone
OAT	-	Outside Air Temperature
P	-	Heating Value of Fuel
Q	-	Heat Flow - Btu/hr.
R	-	Thermal Resistance = sq. ft. x °F x hr./Btu
RAT	-	Room Air Temperature
RSM	-	Rating and Swing Multiplier
SC	-	Shading Coefficient
SEER	-	Seasonal Energy Efficiency Ratio
sq. ft.	-	Square Foot
T.D.	-	Temperature Difference °F
TON	-	12,000 Btu/hr.

INDEX

Z

Zone